ACQUIRING GENOMES

ACQUIRING GENOMES

A THEORY *of the* ORIGINS *of* SPECIES

L YNN M ARGULIS AND D ORION S AGAN

BASIC
BOOKS

A Member of the Perseus Books Group

Hardback edition published by Basic Books, A Member of the Perseus Books Group, in 2002.

First paperback edition published by Basic Books in 2003.

Library of Congress Cataloging-in-Publication Data

Margulis, Lynn, 1938-
    Acquiring genomes : a theory of the origins of species / Lynn Margulis and Dorion Sagan.—1st ed.
        p. cm.
    Includes bibiliographical references
    ISBN 0-465-04391-7 (hc.); ISBN 0-465-04392-5 (pbk.)
    1. Species. 2. Symbiogenesis. 3. Evolution (Biology). 4. Sagan, Dorion, 1959- II. Title.
    QH380 .M37 2002
    576.8'6—dc21

                                                                    2002001521

Text design by *Trish Wilkinson*
Set in 12.5-point AGaramond by The Perseus Books Group

FIRST EDITION

03 04 05 / 10 9 8 7 6 5 4 3 2 1

CONTENTS

TABLES AND FIGURES

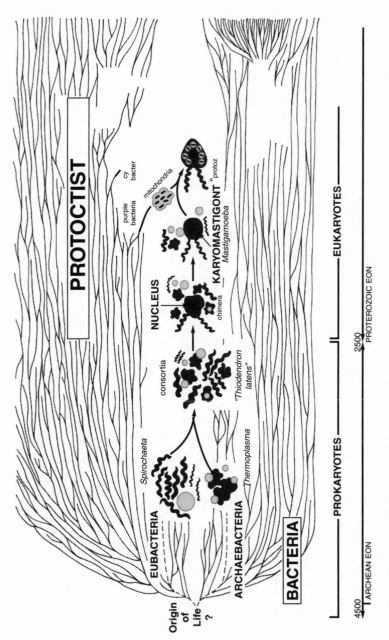

FIGURE FM.1   Bacterial Origin of Nucleated Cells

Any one whose disposition leads him to attach more weight to unexplained difficulties than to the explanation of a certain number of facts will certainly reject my theory.

Charles Darwin
The Origin of Species
1859

FOREWORD

When I got my degree at the University of Berlin, almost eighty years ago, biology consisted of two branches, zoology and botany. What dealt with animals was zoology, and everything else, including fungi and bacteria, was assigned to botany. Things have improved since then, particularly since the discovery of the usefulness of yeast and bacteria for molecular studies. Most of these studies, however, strengthened the reductionist approach and thus fostered a neglect of the major actors in evolution—individuals, populations, species, and their interactions.

The authors of *Acquiring Genomes* counter this tendency by showing the overwhelming importance of interactions between individuals of different species. Much advance in evolution is due to the establishment of consortia between two organisms with entirely different genomes. Ecologists have barely begun to describe these interactions.

Among the millions of possible interactions (including parasitism), the authors have selected one as the principal object of their book: symbiosis. This is the name for mutual interaction involving physical association between "differently named organisms." The classical examples of symbiosis are the lichens, in which a fungus is associated with an alga or a cyanobacterium. At first considered quite exceptional, symbiosis was eventually discovered to be almost universal. The microbes that live in a special stomach of the cow, for instance, and provide the enzymes for its digestion of cellulose are symbionts of the cow. Lynn Margulis has long been a leading student of symbiosis. She convinced the cytologists that mitochondria are symbionts in both plant and animal cells, as are chloroplasts in plant cells. The establishment of a new form from such symbiosis is known as symbiogenesis.

For many years, Margulis has been a leader in the interpretation of evolutionary entities as the products of symbiogenesis. The most startling (and, for some people, still unbelievable) such event was the origin of the eukaryotes by the fusion of an archaebacterium with some eubacteria. Both partners contributed important physiological capacities, from which ensued the great evolutionary success of the eukaryotes—the cells from which all animals, plants, and fungi are made.

Symbiogenesis is the major theme of this book. The authors show convincingly that an unexpectedly large proportion of evolutionary lineages had their origins in symbiogenesis. In these cases a combination of two totally different genomes form a symbiotic consortium which becomes the target of selection as a single entity. By the mutual stability of the relationship, symbiosis differs from other cases of interaction such as carnivory, herbivory, and parasitism.

The acquisition of a new genome may be as instantaneous as a chromosomal event that leads to polyploidy. The authors lead one to suggest that such an event might be in conflict with Darwin's principle of gradual evolution. Actually, the incorporation of a new genome is probably a very slow process extending over very many generations. But even if instantaneous, it will not be any more saltational than any event leading to polyploidy.

The authors refer to the act of symbiogenesis as an instance of speciation. Some of their statements might lead an uninformed reader to the erroneous conclusion that speciation is always due to symbiogenesis. This is not the case. Speciation—the multiplication of species—and symbiogenesis are two independent, superimposed processes. There is no indication that any of the 10,000 species of birds or the 4,500 species of mammals originated by symbiogenesis.

Another of the authors' evolutionary interpretations is vulnerable as well. They suggest that the incorporation of new genomes in cases of symbiogenesis restores the validity of the time-honored principle of inheritance of acquired characters (what is called "Lamarckian inheritance"). This is not true. The two processes are entirely different. Lamarckian inheritance is the inheritance of modified phenotypes, while symbiogenesis involves the inheritance of incorporated parts of genomes.

Perhaps the greatest merit of this book is that it introduces the reader to the fascinating world of the microbes. It cannot be denied that the average student of animals and plants knows little (almost nothing!) about this world. The authors do not limit their account to symbiogenesis but also provide an enthralling description of protists and bacteria. What biologist realizes that animals have only a single set of metabolic pathways while different kinds of bacteria have more than twenty fundamentally different ones? While thousands of specialists study the physiology of vertebrates, the far more diverse physiologies of the bacteria are the province of just a few scores of microbiologists.

Chapters 9 through 11 are devoted to fascinating accounts of dozens of unusual consortia of plants and animals with microbes (protists, lower fungi, and bacteria). It is quite admirable how natural selection succeeded in fusing the most unlikely combinations into single consortia. No whodunit could be more spellbinding than some of these cases of symbiogenesis.

Given the authors' dedication to their special field, it is not surprising that they sometimes arrive at interpretations others of us find arguable. Let the readers ignore those that are clearly in conflict with the findings of modern biology. Let him concentrate instead on the authors' brilliant new interpretations and be thankful that they have called our attention to worlds of life that, despite their importance in the household of nature, are consistently neglected by most biologists.

Let us never forget the important lesson taught by these authors: The world of life not only consists of independent species, but every individual of most species is actually a consortium of several species. The relations between larger organisms and microbes are infinite in number and in most cases make an indispensable contribution to both partners' fitness. Some knowledge of this vast branch of biology should be an essential component of the education of every biologist. There is more to biology than rats, *Drosophila*, *Caenorhabditis*, and *E. coli*. A study of symbiogenesis can't help but lead to a deeper understanding of the world of life, and there is no better way to gain this knowledge than to study *Acquiring Genomes*.

Ernst Mayr
Bedford, Massachusetts
February 2002

*A*mid all the recent interest in complexity, many point out that the future of science belongs more to biology, the study of complex systems, than to physics. Few beings, reader, are more complex than you. In this book we argue that to understand the true complexity of life—the main source of evolutionary novelty Charles Darwin sought—one must understand how organisms come together in new and fascinating ways, and how their genes are donated and acquired. This then will be the story of how species, and speciation itself, evolved through the acquisition of genomes. With more than a century of observation and experimentation by scientists around the world, and intense communication among them, let us now explain how new species come into being.

Neither an omnipotent deity nor fantastic good luck enters into it. The story that begins with Jean Baptiste Lamarck's studiousness, Charles Darwin's data, Gregor Mendel's garden peas, James Watson's willfulness, and Francis Crick's lack of humility ends, with a species of romantic irony, in the muck and slime. Its protagonists are speedy, determined bacteria and expert protist architects on a tectonically active Earth under an energetic sun. Wars, alliances, bizarre sexual encounters, mergers, truces, and victories are the dramas. Random DNA mutations, primarily destructive in their effects, account only for the beginnings. The role of randomness has been exaggerated in the evolutionary saga. Drugs, chemical compounds, and other molecules are mere stimulants and components. Live beings by contrast are the protagonists.

The DNA molecule, like the computer disk, stores evolutionary information but does not create it. Selfish genes, since they are not "selves" in any coherent sense, may be taken as figments of an overactive, primarily English-speaking imagination. The living cell is the true self. An entity that cannot help creating more copies of itself, it plays center stage. The engine of evolution is driven by tiny selves of which we are only half conscious. We fear and malign them. The bacteria, these lively minute beings, escape our awareness except in freakish moments when they alarm or threaten us. We ignore or disdain the fact that they have lives of their own. But they do. The actions of the bacteria and other subvisible selves perpetuate old and generate new species.

On Saturday and every other night in the microbial world, something dramatic goes on. The spectacle has been playing for at least 3,000 million years, and the first acts are far too important for us to miss. Only at the very end of this long history does the social ape who walks upright and gossips nonstop take center stage. The ape who is nearly hairless, the one who cuts down rain forest trees with abandon, is deluded by visions of his importance. His sense of uniqueness and selfish interests combined with his technological profligacy have led to an unprecedented population explosion of primates. But this story is not about that ape, his lovemaking, his cohabitors, or his victims. Rather it is the tale of the nonhuman ancestors that preceded him. Before campfires, before proclamations of independence, before cities and urban sprawl, the Earth around the sun was populated by innumerable kinds of superficially alien life. The whole evolutionary saga of how species originate and how they extinguish may be the greatest tale ever to be told. It is everybody's story.

PART ONE

THE EVOLUTIONARY IMPERATIVE

CHAPTER 1

DARWINISM NOT NEODARWINISM

*C*harles Darwin's landmark book *The Origin of Species*, which presented to scientists and the lay public alike overwhelming evidence for the theory of natural selection, ironically never explains where new species come from.

Species are names given to extremely similar organisms, whether animals, plants, fungi, or microorganisms. Because we need to identify poisons, predators, shelter materials, fuel, food, and other necessities, we have long bestowed names on living and once-living objects. Species names of organisms you probably know are listed in Table 1.1. Until the Renaissance, however, names of live beings varied from place to place and were seldom precisely defined. The confusion of local names and inconsistent descriptions led the Swedish naturalist Carolus von Linné (1709–1789) to bring rigor and international comprehensibility to the descriptions. Since Linnaeus (his Latin name) imposed order on some 10,000 species of live beings, scientists use a first name (the genus—the larger, more inclusive group) and a second name (the species—the smaller, less inclusive group) to refer to either live or fossil organisms.

Most Linnaean names are Latin or Greek. By today's rules the species and genus names are introduced into the scientific literature with a "diagnosis," which is a brief description of salient properties of the organism: its size, shape, and other aspects of its body (its morphology); its habitat and way of life; and what it has in common with other members of its genus. The diagnosis appears in a published scientific paper that describes the organism to science for the first time. The paper also includes details beyond the diagnosis, called the "description." To be a valid name not only must the names, diagnosis, and description be published but a sample of the body of the organism itself must be deposited in a natural history museum, culture collection, herbarium, or other acknowledged repository of biological specimens.

Fossils are dead remains, evidence of former life. The word comes from *fosse* in French, something dug up from the ground. Fossil species, like the enigmatic trilobite *Paradoxides paradoxissimus*, also are given names and grouped on the basis of morphological similarities and differences.

The word "species" comes from the Latin word *speculare,* to see—like spectacles or special. Everyone, knowingly or not, uses the morphological concept of species—dogs look like dogs, they are dogs, they are all classified as *Canus familiaris.* The problems come when we try to name coyotes (*Canus latrans*), wolves (*Canus lupus,* gray wolf, or *Canus rufus,* red wolf), and other closely related animals.

Zoologists, those who professionally study animals, have imposed a distinct concept of species, which they call the "biological species concept." Coyotes and dogs in nature do not mate to produce fully fertile offspring. They are "reproductively isolated." The zoological definition of species refers to organisms that can hybridize—that can mate and produce fertile offspring. Thus organisms that interbreed

Table 1.1—Some Familiar Species

Common Name	Genus	Species
*E. coli** colon bacteria	*Escherichia*	*coli*
corn	*Zea*	*mays*
dogs	*Canus*	*familiaris*
fruit flies	*Drosophila*	*melanogaster*
green mold that makes penicillin	*Penicillium*	*chrysogenum*
oranges (orange trees)	*Citrus*	*aurantiacus*
poison ivy	*Rhus*	*toxicum*
people	*Homo*	*sapiens*
pond amebas	*Amoeba*	*proteus*

* Here, because of media-fed concern over the dangers of becoming sick from improperly cooked food, the Latin name has worked its way back into popular culture.

(like people, or like bulls and cows) belong to the same species. Botanists, who study plants, also find this definition useful.

A third concept of species is in vogue today: the "phylogenetic," "evolutionary," or "cladistic" species idea. Groups of organisms, again like people or corn plants or chickens, considered to be all descended from the same ancestors ("clade") are classified as members of the same species. Such organisms are called "monophyletic" because they are descended from "a single common ancestor."

We all have a strong sense of species—our ancestors needed it to recognize food, potential mates, thatch grass, poisonous snakes, and many other organisms in order to survive. Two-year-old children delight in recognizing domestic animals, birds, and even fish; witness the popularity of stuffed teddy bears and dinosaurs. An instinctive evolutionary cognition of life forms has been crucial to our and other species' survival.

Evolution, the study of changes of life through time, is largely the tracking of the origins of species. We argue here that of these three concepts of species, the traditional, morphological one is still the best. The morphological species, we will show, is the external manifestation of the symbiogenetic species.

The assignment of similar animals of compatible genders to the same species if, in their natural habitat, they can mate and produce viable offspring is adequate for mammals and many other animals but it is not general. Here we widen the concept of speciation to include all organisms. Our symbiogenetic definition of species is as follows. We suggest that if organism A belongs to the same species as organism B, then both are composed of the same set of integrated genomes, both qualitatively and quantitatively. All organisms that can be assigned to a unique species are products of symbiogenesis. That is, because A and B share the same number of the same different kinds of integrated genomes they are assigned to the same species. Since no bacterium (whether eubacterium or archaebacterium) evolved from symbiotic integration of formerly independent cells, bacteria lack species; the process of speciation began with the earliest eukaryotes (the first protists, or organisms with nuclei). The concept that all bacteria are interfertile (they can transfer their genes from one to another no matter how different are the recombining partners) has been well argued for over thirty years and is newly explained in *Prokaryotology*, by Sonea and Matthieu, 2000. Ironically the popular evolutionist's view that organisms evolve by the accumulation of random mutation best describes the evolutionary process in bacteria. All of the larger, more familiar organisms originated by symbiont integration that led to permanent associations. The once-separate symbiotic components become genetically integrated to make new whole individuals, always in populations. As we see from the work on karyotypic fissioning in Chapter 12, the ancient microbial symbionts became so stripped-down in their capabilities and morphology that their true nature can be revealed only by fervent sleuthing. Whatever the origin of the evolutionary variation under study, it is natural selection that relentlessly eliminates the beings whose form, physiology, behavior, and chemistry are not suited for that given environment at that given time and place (whatever the details). Bacterial cells have single genomes that acquired their sets of genes, usually a thousand or more, from compatible prokaryotes one or two at a time. Eukaryotes acquire and integrate entire complete genomes to form "individuals." For example, all plant cells have at least four integrated genomes: 1) the motile eubacterium and 2) the protein-synthesizing archaebacterium that formed the nucleated cell (the first protist) followed by 3) the oxygen-respiring proteobacterium that became the mitochondrion and finally 4) the cyanobacterium that became the chloroplast.

The biologists' "interbreeding" requirement (the "biological species definition") is extremely useful but mostly for land-dwelling mammals, other closely related forms like birds and snakes, and many plants. The requirement for potential mating (hybridization) is probably related to the way the animals have evolved by karyotypic fissioning, the subject of Chapter 12. But it does not apply to at least four-fifths of all life. The biological concept of species should be renamed the "zoological-botanical concept of species." In zoology the concept is, indeed, indispensable, and in botany it is useful.

However, the "phylogenetic" or "evolutionary" or "cladistic" concept of species is entirely wrongheaded, and its adoption interferes with understanding how species arise. The long-term symbiosis that led to species origin by symbiogenesis requires integration of at least two differently named organisms. No visible organism or group of organisms is descended "from a single common ancestor."

The purpose of this book is to explain, with abundant evidence collected by scientists around the world, this new concept of how new species really come into being.

DARWIN'S EVOLUTION

"Evolution" is a word Charles Darwin never used. Rather he wrote about "descent with modification." Even so, the basic modern concept of evolution is undoubtedly attributable to Darwin. If we raise our hands to swear that we are darwinists, we also swear, with equal

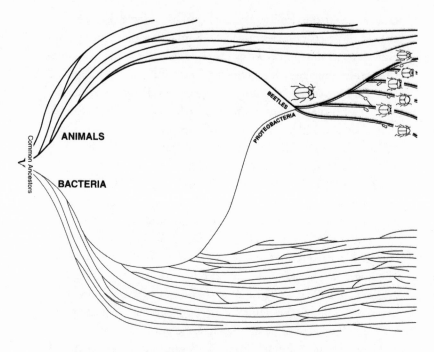

FIGURE 1.1   Subvisible Anastomosis in Beetle Phylogeny

fervor, that we do not agree with his followers the neodarwinists and other modern evolutionists. Darwin's original views must be distinguished from those of his successors. Everyone today who studies modern biology, indeed virtually every scientifically minded modern person, is a Darwinist. But since neodarwinism and "the modern synthesis" that attributed evolutionary change to random mutation developed between 1930 and 1960, long after Darwin's death (in 1882), he himself never even knew about it. Nor did he know about "evolutionary biology." The "modern synthesis" was the science invented to unite Darwin's idea that organisms and populations gradually change with Gregor Mendel's concept of genetic stasis.

What are the differences between darwinism and prevalent modern ideas? And how do the concepts of this book differ from the ideas of today's neodarwinists?

Let us start with the ideas with which we all concur. Darwin recognized that all organisms, whether domesticated animals, earthworms whose "castings" (fecal excretions) produce "vegetable mould," or barnacles (all of which he studied), all the time produce more offspring than can survive. More are born, hatched, budded, and produced in some way or other than ever can grow to maturity. Only a fraction—sometimes a tiny fraction like 1/100,000, sometimes as many as 50 percent, but always only a fraction—of potential organisms ever have offspring that survive to have their own offspring. Only a minute minority continue to produce descendants. Darwin recognized this fact and called the idea "natural selection." Fundamentally, "natural selection" is a way to express the concept that most life does not persist through time. "Differential survival" is all it really is. One may phrase it like this: Populations of all species potentially grow exponentially. In numbers of offspring per generation (or per unit time) the tendency to population growth is called "biotic potential," the maximal number of offspring possible for a given species. Natural selection refers to the fact that biotic potential is seldom reached. The biotic potential of people is about twenty children per couple generation (or twenty children per twenty-five years). The biotic potential of the woodland fungus *Alternaria* sp. is about 10^5 (one hundred thousand) spores per minute for six months. That of dachshunds is eleven pups/litter at three litters/year for ten years, a total of 333 pups per couple per generation. And so on. Biotic potential is species-specific and can be expressed as the greatest possible number of offspring born, hatched, budded, germinated from seed, or otherwise produced per generation.

Darwin saw clearly that offspring organisms differ from each other, and some of these differences—those, he claimed, that are "important to us"—are inherited. So he put together these two ideas: variations occur, and only some offspring survive to produce their own offspring. Differences among organisms are selected by the environment through differential survival.

Thus he came to realize that organisms are connected through time and all descend from common ancestors. He called his idea descent with modification. Some of his best examples came from dogs and pigeons. He observed that people who bred chickens for eating selected those with bigger breast muscles, and the people who raised dogs for hunting selected those with the keenest senses. He noticed that selection relatively quickly produced differences in offspring, which in dogs, for instance, were codified as different breeds.

Darwin's basic concept, as modified, extended, and verified in the 20th Century, has been summarized by Ernst Mayr (2001) in his book *What Evolution Is*. This accessible introduction to the grand sweep of evolutionary science by an active participant expands on our brief resumé here.

In short, Darwin's argument goes: abundant production of too many organisms where only some survive to reproduce can always be shown. All offspring vary from their parents. Some of these variations are inherited. Therefore changes occur in the history of life. But what puzzled Darwin was, where does the inherited variation come from? Why, that is, aren't organisms always just like their parents, or just like their single parent? What is the source of evolutionary variation? In the end, Darwin didn't know. But he did note that much variation is never inherited: circumcised male children, most tailless mice, and yellowed leaves did not give rise to foreskinless babies, tailless mice, or plants with yellow leaves. Darwin often wrote that only variation that is inherited is important for "descent with modification," that is, evolution. The length of one's hair, as a variation, would not be of interest, because it is not inherited. But the length that hair might grow if permitted to grow maximally might be of interest as an example of inherited variation.

In summary, any population displays measurable variation, much of which is irrelevant to evolution. But the same population also shows variation that is inherited and therefore relevant to evolutionary change. Such variation can be detected and measured in any group. So our question becomes the one Darwin asked himself—what is the source of observable inherited variation?

SYMBIOSIS AS A SOURCE OF INHERITED VARIATION

The word among neodarwinists, primarily zoologists who today call themselves "evolutionary biologists," is that inherited variation derives from random changes in the chemistry of the genes. Variations that are heritable are caused by mutations, and mutations are random. Unpredictable, and uncorrelated with behavior, social conditions, food, or anything else, mutations are permanent genetic changes. As these random genetic changes accumulate with time, they determine the course of evolution. Such is the view offered by most evolutionary literature.

We certainly agree that random heritable changes, or gene mutations, occur. We also concur that these random mutations are expressed in the chemistry of the living organism. Altered proteins that can be traced back to gene mutations in living organisms have been massively demonstrated. The major difference between our view and the standard neodarwinist doctrine today concerns the importance of random mutation in evolution. We believe random mutation is wildly overemphasized as a source of hereditary variation. Mutations, genetic changes in live organisms, are inducible; this can be done by X-ray radiation or by addition of mutagenic chemicals to food. Many ways to induce mutations are known but none lead to new organisms. Mutation accumulation does not lead to new species or even to new organs or new tissues. If the egg and a batch of sperm of a mammal is subjected to mutation, yes, hereditary changes occur, but as was pointed out very early by Hermann J. Muller (1890–1967), the Nobel prizewinner who showed X-rays to be mutagenic in fruit flies, 99.9 percent of the mutations are deleterious. Even professional evolutionary biologists are hard put to find mutations, experimentally induced or spontaneous, that lead in a positive way to evolutionary change.

We show here that the major source of inherited variation is not random mutation. Rather the important transmitted variation that leads to evolutionary novelty comes from the acquisition of genomes. Entire sets of genes, indeed whole organisms each with its own genome, are acquired and incorporated by others. The most common route of genome acquisition, furthermore, is by the process known as symbiogenesis.

What is symbiogenesis and how is it related to symbiosis? First, what is symbiosis? Symbiosis is simply the living together of organisms that are different from each other. When originally defined by Heinrich Anton de Bary (1831–1888), symbiosis was the living together of "differently named organisms." Symbioses are long-term physical associations. Different types of organisms stick together and fuse to make a third kind of organism. The fusion is not random. Symbiotic relationships occur under specific environmental conditions. In some of these relationships, one partner in the symbiosis feeds off the other to its detriment and even death. Such exploitative associations are called "parasitic" or "pathogenic." They tend to be highly sensitive to environmental stress. The parasite that invariably and virulently kills its partner kills itself. With time and circumstance the nature of associations tends to change. The relationships that interest us most here are modulated coexistence between former predators, pathogens and their hosts, their shelter and food sources. As members of two species respond over time to each other's presence, exploitative relationships may eventually become convivial to the point where neither organism exists without the other.

Long-term stable symbiosis that leads to evolutionary change is called "symbiogenesis." These mergers, long-term biological fusions beginning as symbiosis, are the engine of species evolution. A very specific example of symbiogenesis in live organisms illustrates the point. We introduce you to photosynthetic animals, actual plant-animal hybrids. Slugs, the familiar shell-less mollusks eating your garden plants, have entirely green photosynthetic relatives. The ancestors of these slugs have eaten but not digested certain green algae, which years ago entered the tissues of the animal—and stayed there. All members of these species (for example *Elysia viridis*) are always green. These underwater slugs need not seek food. Rather they crawl near the shore. They never eat throughout their adult life. The slugs, newly evolved green animals, now sunbathe in the way plants sunbathe. They are permanently and discontinuously different from their gray, algae-eating ancestors. Such acquisition, use, and permanent inheritance of entire alga genomes only seems marginal—in fact it has occurred many times in evolutionary history. At least four or five times different lineages of green animals have been documented in videos and scientific papers. Green animals provide graphic examples of symbioses that lead to symbiogenesis.

A second example is lichen, the plantlike greenery that grows on rock and tree bark. Most biologists do not realize that they are Schwendenerists. A Schwendenerist is a scientist, usually a botanist, who believes that lichens are not plants. The Schwendenerism argument raged in the late nineteenth century. Simon Schwendener (1829–1919) explained, in a long, complex treatise written in German, the composite nature of the plantlike lichen-growths and why no lichen is a plant. Modern investigators all accept lichens as symbiogenetic in origin. One quarter of all documented fungi are now known to be "lichenized"—they must live photosynthetically with green algal or cyanobacterial partners. This accounts for some 25,000 species of lichens.* Together the intertwined organisms act as units: They produce highly organized and structured tissue–stalks, leaves, and bulbous protrusions. The fungal and algal (or cyanobacterial) cells of lichens propagate together. No serious biologist today ever suggests that any lichen is a true plant. All concur that while lichens look like plants and photosynthesize as plants do, even

---

* Estimated; only 13,000 have been formally introduced into the scientific literature. (Brodo, et al., 2000)

cursory morphological analysis shows they differ from plants fundamentally.

Lichens provide us with the classic example of symbiogenesis. Moreover, the lichen individual is something distinct from either of its two component parts. It is neither the green alga (or cyanobacterium) nor a fungus. It is lichen. Lichens, evolutionary novelties that evolved by cyanobacteria or algal genome acquisitions, have taken a unique path and display characteristics distinct from both sets of their ancestors. Though traditionally studied within botany, lichens have always been central to concepts of symbiosis and symbiogenesis in evolutionary thought. And yet their symbiotic nature has led them to be thought of as a marginal evolutionary phenomenon. Perhaps they have been accepted as an example of the power of symbiogenesis to generate evolutionary novelty because the partners are of the same size. The alga and fungus are both easily seen with low-power microscopy, so neither can be studied without simultaneous study of the other. In some green animals, by contrast, for example in the flatworm species *Convoluta roscoffensis*, the two partners differ greatly in size. The worm is large, centimeters, whereas the little photosynthetic microorganisms, the algae, are microscopic. Such size discrepancies make the symbiosis, and symbiogenesis, less obvious.

The usual attitude is, "Well, symbiosis is acceptable for the evolution of microbial things that interest you" (as if we had a kind of bizarre personal issue with microorganisms). "But we don't believe symbiosis is an important evolutionary mechanism for 'higher' organisms (mainly mammals) that really interest us." But this book is replete with examples where symbiogenesis serves as the source of evolutionary novelty in familiar animals. For example, cows. Cows are "forty-gallon fermentation tanks on four legs," according to Sir David C. Smith (Smith and Douglas, 1987). Cows ingest grass, but they never digest it because they are incapable of cellulose breakdown. Digestion in cows is by microbial symbionts in the rumen. The rumen is a special stomach, really an overgrown esophagus, that has changed over evolutionary time. Cows that lack rumens don't exist; cows (and bulls) deprived of their microbial symbionts are dead.

Random DNA base changes of course play a role in the evolutionary process. They are like printers' errors that crop up in the copying of books. They rarely clarify or enhance meaning. Such small random changes are nearly always inconsequential or detrimental to the work as a whole. We do not deny the importance of mutations. Rather we insist that random mutation, a small part of the evolutionary saga, has been dogmatically overemphasized. The much larger part of the story of evolutionary innovation, the symbiotic joining of organisms (similar, if we extend the printing analogy, to the fusion of texts in plagiarism or anthologies) from different lineages, has systematically been ignored by self-proclaimed evolutionary biologists (Sapp, 2002).

In the familiar phylogenetic tree, the acquisition of heritable genomes can be depicted as an anastomosis, a fusing of branches. The major proposition put forth here, of fusion of evolutionary lineages, is sometimes decried as an alternative to classical darwinism. But symbiogenetic acquisition of new traits by inheritance of acquired genomes is rather an extension, a refinement, and amplification of Darwin's idea. Such evolution requires new thought processes. New metaphors to reflect on permanent associations are needed. Symbiosis, merger, body fusion, and the like cannot be reduced to replacing "competition" as a major motive force in evolution with "cooperation." Ultimately, an anthroprocentric term like "competition" has no obvious place in the scientific dialogue. Rather we would propose a new search in the social sciences for terms to replace the old, tired social darwinist metaphors. If survival is owed to symbiosis, rather than overemphasized intraspecific competitive struggles, what then are the consequences for nonbiologists interested in evolution?

In this book we have been careful to never use either "cooperation" or "competition" to describe biological or other evolutionary phenomena. These words may be appropriate for the basketball court, computer industry, and financial institutions, but they paint with too broad a brush. Far too often they miss the complex interactions of live beings, organisms who cohabit. Competition implies an agreement, a set of actions that follow rules, but in the game of real life the "rules"—based on chemistry and environmental conditions—change with the players. To compete—for example on opposite teams—people must basically cooperate in some way. "Competition" is a term with limited scientific meaning, usually without reference to units by which it can be measured. How does the green worm or the lichen fungus assess its competitive status? By the addition of points in its score or by dollars or Swiss francs? No. Then what are the units of competition? If you ask what are the units of biomass we can tell you in grams or ounces. If you ask how light or biotic potential is to be measured, we answer in lux or foot-candles or number of offspring per generation. But if you ask "what are the units of competition" we reply that yours is not a scientific notion. Vogue terms like "competition," "cooperation," "mutualism," "mutual benefit," "energy costs," and "competitive advantage" have been borrowed from human enterprises and forced on science from politics, business, and social thought. The entire panoply of neodarwinist terminology reflects a philosophical error, a twentieth-century example of a phenomenon aptly named by Alfred North Whitehead: "the fallacy of misplaced concreteness." The terminology of most modern evolutionists is not only fallacious but dangerously so, because it leads people to think they know about the evolution of life when in fact they are confused and baffled. The "selfish gene" provides a fine example. What is Richard Dawkins's selfish gene? A gene is never a self to begin with. A gene alone is only a piece of DNA long enough to have a function. The gene by itself can be flushed down the sink; even if preserved in a freezer or a salt solution the isolated gene has no activity whatsoever. There is no life in a gene. There is no self. A gene never fits the minimal criterion of self, of a living system. The time has come in serious biology to abandon words like competition, cooperation, and selfish genes and replace them with meaningful terms such as metabolic modes (chemoautotrophy, photosynthesis), ecological relations (epibiont, pollinator), and measurable quantities (light, heat, mechanical force). So many current evolutionary metaphors are superficial dichotomizations that come from false clarities of language. They do not beget but preclude scientific understanding.

Would not society be better served, then, if we adopted symbiotic metaphors instead of competitive ones? No. Society will be better served by more accurate scientific understanding, and this is not to be gained by substituting one pole of oversimplified metaphors for another. But of course organisms do vie in various ways with each other for space and food. Such vying however (or competition) among members of the same species does not in itself lead to new species; a source of genetic novelty—usually symbiogenesis—is needed. Only a small fraction of any population survives. Is this not then competition? Is it not then a "struggle for existence"? Not really. When Darwin referred to struggle, he meant the tendency of all organisms to grow, to reproduce, and to attempt to leave their own descendants. He did not mean that God battles with angels or that fistfights ensue among the mistresses of the king. In the sense of the tendency of some but not all to leave offspring, Darwin's proper term is natural selection. The "struggle" is the bald fact, a rule of all life, that biotic potential is never reached. Only a few of us produce offspring who go on to produce offspring who themselves produce more fertile offspring. To call the tendency to leave offspring or fail to do so "competition," as biologists frequently do, is misguided.

As a highly social species so often concerned with relationships among ourselves, we tend to be oblivious to our relationships with other species. Biologically, for example, there is no such thing as a symbiosis between a mother and her unborn infant. Symbiosis is always a physically close relationship between organisms of *different* kinds. We humans have a symbiotic relationship with our eyelash mites. Most of us ignore the fact that we live symbiotically with our eyelash mites or with our underarm or gut bacteria or with the spirochetes in our gum tissue. But we do have eyelash mites and intestinal and underarm bacteria. We are inattentive to our immense populations of oral spirochetes until our gums bleed or our tooth aches. Each of us harbors bacteria in our intestines that make K and B vitamins that are absorbed through the intestinal walls. We seem reluctant to acknowledge our symbiotic bacteria even when we see them with a scanning electron microscope. We all enjoy silent, unconscious relationships with microbes.

But none of us has a symbiotic relationship with his or her mother-in-law, father, or adolescent child. Why? Because all people belong to the same *Homo sapiens* species. "Symbiosis" is an ecological term that describes two or more organisms of different kinds in protracted physical contact. Even bees and flowers, associated as they are by pollination ecology, do not have a symbiotic relationship. They enjoy other kinds of relationships, but they are not in physical contact long enough to be symbionts. Eyelash mites and the human body are symbionts. The odors that your socks emanate come from some of the bacteria and fungi that live between your toes; they have special habitats. The estimate is that any person is about ten percent dry weight symbioses. Most human symbionts, by weight, are the many types of bacteria that thrive in the colon (the large intestine). Our relation to these microbes is one of association—not "benefits" or "costs" or "cooperation" or "competition." Symbiosis and its evolutionary consequence, in some cases "symbiogenesis," is simply a fact of life.

In short, much of the modern evolutionist's (not the quaint darwinist's) terminology should be abandoned. Both in popular culture and especially in the scientific "evolutionary biology" subculture, the terminology is not scientific, but misleading to the point of being destructive. Honest critics of the evolutionary way of thinking who have emphasized problems with biologists' dogma and their undefinable terms are often dismissed as if they were Christian fundamentalist zealots or racial bigots. But the part of this book's thesis that insists that such terminology (Table 1.2) interferes with real science requires an open and thoughtful debate about the reality of the claims made by zoocentric evolutionists.

THE MYTH OF THE
INDEPENDENT INDIVIDUAL

Perhaps you are suspicious of the suggestion that organisms aren't as self-contained as we thought. The idea that we people are really walking assemblages, beings who have integrated various other kinds of organisms—that each of us is a sort of loose committee—opens up too many challenging speculations. When "the committee" gets sick, is simply a single animal getting sick, or is illness more a re-arrangement of the members? We imagine that pathogenic microbes attack us, but if such pathogens are part of the committee that makes up each of us to begin with, isn't health less a question of resistance to invasion from the outside and much more an issue of ecological relationships among committee members? Yes.

We humans, like all organisms, live embedded in ecological communities. If, as individuals, we feel we are falling apart, it is probably because we multicomponented beings *are*, in fact, falling apart. Each person, each dog, each tree is composed of many different living parts that can be detected and identified. The relations among our living component parts are absolutely critical to our health, and therefore to our happiness. The completely self-contained "individual" is a myth that needs to be replaced with a more flexible description. The symbionts of people are difficult to study, however, for many reasons: complexity of disparate sizes, inability to carry out experiments in human heredity, politics, and social prejudice. Lichens provide better cases for understanding

Table 1.2—Evolutionary Terms as Battle Cries*

(No adequate quantitative measure of these exists; therefore they are deficient, even pseudoscientific terms.)

altruism	indirect fitness
benefit	levels of selection
game payoff matrix	lower organism
gene, selfish gene	mate competition
group selection	mutualism
higher organism	parental investment
inclusive fitness	reciprocal altruism

* Only a few of many examples listed here. See for example E. F. Keller and E. A. Lloyd, 1992, and L. Keller, 1999, for wholesale acceptance of neodarwinist "holy writ." See Margulis, 1990, for spirited criticism against this fallacious language of misplaced concreteness. See page 26.

symbionts. If certain lichens are placed in the dark, the photosynthetic member, the photobiont (usually a cyanobacterium such as *Nostoc* or a green alga such as *Trebouxia*), cannot live. The fungus often just grows and grows—it digests its green former partner. If lichens are placed underwater for a long time and in the light, the fungus drowns but the green alga will just grow and grow. Lichens, therefore, are composite organisms that require mixed light: They do not survive persistent *all*-light or *all*-dark conditions. So, too, they cannot live when the environment is entirely wet or entirely dry but require cycling between extremes. Change in the environment is essential to their survival. This lichen proclivity for change should not trouble us. The cycles, the alternations between wet and dry and light and dark, are what maintain the living composite, the apparent individual. Certain ratios of changes are required for most living beings to persist and propagate.

We must begin to think of organisms as communities, as collectives. And communities are ecological entities.

To go beyond animals, think of plants. We stick their seeds or seedlings in the garden soil and marvel as plants do their thing.

Most plant roots live in the rhizosphere. This is an ecological zone of many different organisms that all grow and metabolize at the same time. Some rhizosphere inhabitants provide nutrients to what we see as the plant, generally the part above the ground. But, like animals, plants are also confederacies of once-separate and different kinds of organisms. As we will see, symbioses in the roots, in the leaves, and even on the stems are known to have produced new forms of plant life—and may be responsible for the origin of those once-monstrous growths without which humanity as we know it would never have evolved—fruits.

The book *What Is Life?* (Margulis and Sagan 2000) has a photo taken from a distance by Connie Barlow that shows a stand of poplar trees in Colorado. Anyone can count these trees—there are hundreds if not thousands of them. But although it has many parts, this stand is really only one single organism. Under the ground the "tree" is continuous. It forms a connected structure with many upright shoots that emerge from the soil, straining our everyday notion of a single tree. The "individual," whose roots are completely continuous, extends for kilometers laterally and for meters into the ground and up into the air vertically. This poplar stand is believed to be one of the largest "organisms" alive today.

Plants must be integrally incorporated into our conception of the evolutionary process. As stated, a problem with modern "evolutionary biologists" is that their examples are nearly always derived from people or other animals, especially other land mammals. Occasionally fruit flies or other insects serve as their illustrations of evolution. When they say "lower organisms," they are generally speaking of animals other than mammals. As zoologists they tend not to know the microbial world—they are often ignorant of bacteria, fungi, and the myriad other "larger" microbes called protists. Zoologists tend to study very little botany, very little protistology, and no bacteriology. They write about "individuals," but what is generally meant is people, pets, and our zoo and food animals. Occasionally, modern evolutionists factor agricultural plants into their analyses, but they often do so in a limited and scientifically inadequate way. Although zoocentrism may be adequate for the kinds of mammals that are deployed for breeding further populations of mammals, it is a tame approximation—a kind of Apollonian hallucination—of what is going on with life as a whole.

A nineteenth-century question still with us today is whether evolutionary progress exists. Is increasing complexity on a large evolutionary scale to be understood as progress?

Evidence for evolutionary expansion is easy to show. In the fossil record evidence for the outward expansions of life forms abounds. Life, of course including human life, loves to be where water, ocean or lake, contacts the air as well as the soil. We all know from real estate values that shorelines are popular environments. Life enjoys habitats where water meets land meets air. Most life forms within a few millimeters of such surfaces.

Life apparently evolved from propagation at the seaside and only afterward expanded into the polar regions, high montagne lakes, and ocean abyss. Only since the Cenozoic era, which began 64 million years ago, did animal and algal life forms leave body fossils in the high Arctic and Antarctic. (For the geological time scale, simplified, see Table 9.1, p.147.) The continuous core of life on Earth has expanded and extended its range. Is this progress? Examples of fossil reptiles that sported more vertebrae and more morphological complexity than any now living have been unearthed. Since they are extinct, they stand in opposition to the concept of evolutionary progress. The very term "progress," with its moral overtones, denotes a complex quantity that is unmeasurable and unassessible. The descendants of these reptiles lost complexity— they are simplified relative to their ancestors—but we can not say that they "regressed." They evolved to have fewer vertebrae, that's all. That some directional progress in evolution led to us, *Homo sapiens,* on our peak at some Olympian summit is an untenable concept. As more of Earth became covered with more life, life *did expand*, but whether it "progressed" is questionable. Life's apparent progress is best seen in the context of its conformity to the second law of thermodynamics, as we see in Chapter 2. We trace life's history from the Precambrian to the present, always mindful of the question "How do new species evolve?" We recast the concept of "evolutionary progress" and "life's purpose" in terms of the new thermodynamics, which unites and integrates, in a way distinct from but complementary to genetics and molecular biology, the physical and biological sciences.

DARWIN'S DILEMMA

Darwin's book *The Origin of Species by Means of Natural Selection or the Preservation of Favoured Races in the Struggle for Life* sold out (to the bookseller) its 1,200 copies on the day it was published, November 24, 1859. Readers were conscious of buzzing rumors: the author of the popular *Voyage of the Beagle* would explain how man descended from apes, how laughter and grimace were universal facial expressions in slaves and kings, and how the one and only Almighty God had retired from his job as producer, director, and chief executive officer of the known world. The reclusive scholar, who had painstakingly assembled his theory over so many years since his youthful employment as the *Beagle's* naturalist, at last would tell it all. Everyone knew that this Darwin was thoroughly trustworthy and careful and that, since his family fortune and that of his wife, Emma Wedgwood, permitted him leisure to pursue truth, his book ought to be accurate and dependable.

Probably many more bought the book than read it. After a century of exegesis by savants and much translation into common language, the authorities have agreed that Darwin's insights were many and valuable. The book itself, however, is laced with hesitancies, contradictions, and possible prevarication. Darwin made it clear that Man, like all live beings, survived to the present preceded by an immense and daunting history. No God had made either Man or tomato. Nor had any other form of life been created separately in seven or fewer days. Yet Darwin, perhaps mainly for reasons of political acumen, did reserve the possibility that God had begun life in the first place.

Surprisingly, when all was said and done about "grandeur in this view of life" (one of Darwin's last phrases in the great book), it was abundantly clear that in 500 pages of closely spaced type the title question—on the origin of species—had been entirely circumvented—abandoned, ignored, or coyly forgotten. As the Australian biologist George Miklos so appropriately put it:

> The "struggle for existence" has been accepted uncritically for generations by evolutionary biologists with the *Origins of Species* quoted like so much Holy Writ, yet the origin of species was precisely what Darwin's book was not about.

For those who skimmed the book, those who read his myriad other works, and those who simply learned about the book's contents from others, Charles Darwin ultimately established, to the satisfaction of his scientific contemporaries and followers, a major idea entirely valid in our day. All species of life did descend from related predecessors. All life, whether or not made by a deity in the very beginning, is connected back through time to preexisting, proximally similar life forms. Today, with our better understanding of cosmic evolution and the chemistry of life's origins, any requirement for a deity can be pushed back still further, to the mysterious origins of the cosmos in the Big Bang.

Darwin showed clearly how living things "beget" descendants that, inevitably, differ slightly from their parents. He noted that many of these identifiable differences are inherited. Traits can be seen transmitted, often distinguishable but only slightly changed, to puppies, colts, chicks, calves, children, and grandchildren. Darwin publicized the fact that was there for all to see: of the large numbers of offspring that potentially grow from seed, hatch from eggs, or are born out of the womb, only a small number survive to produce their own offspring. Perhaps many, whether acorn or kitten, surround their mothers at first, but few, if the course of their lives are traced, actually survive to bring their own issue to fruitful adulthood. By logic, Darwin showed, the survivors must have traits that are more conducive to survival in that particular environment than offspring that did not survive. As said, Darwin gave this process of differential survival and reproduction the name "natural selection."

"Natural selection" for Darwin did not imply a lugubrious elderly powerful inhabitant of a cloudy sky—there was no "Natural Selector." Instead the expression denoted the survival and reproduction of the chosen few relative to the prodigious many. Darwin conceded that the naturally selective process, by itself, did not seem to create novelty; rather, from the vast store of variants, differing organisms in nature, it only eliminated offspring that already existed by their failure to reproduce. How then did Darwin's intrinsic, inherited variation arise in the first place?

Darwin would have us believe that the entire concept of evolution originated with him. He consistently failed to credit his energetic paternal grandfather, Erasmus Darwin. The contribution of Erasmus, a medical doctor and progressive poet who wrote (in *Zoonomia,* 1794–1796) about evolution by natural selection, was taken as less than serious by his grandson. The first modern naturalist to publish a great body of literature that argued for the evolution of all modern life from ancestral predecessors was Jean Baptiste Lamarck (1744–1829). In English-speaking circles, Lamarck is taught as the Frenchman who made a negative contribution to science with his erroneous claim that characteristics acquired by an animal or plant may be inherited in the descendants of the acquirer.

"Inheritance of acquired characteristics," the phrase inseparable from the name of Lamarck, is taught as equivalent to "Lamarckianism"—and "wrong." But Darwin too, like Lamarck, struggled with the problem of the ultimate source of heritable variation—and came up with wrong answers. That Darwin invented, in the end, a Lamarckian explanation—his "pangenesis" hypothesis to explain how heritable variations arise—tends to be forgotten, as described in Mayr's book (1982). By his reckoning, "gemmules," theoretical particles borne by all living beings and subject to experience during the lifetime of their bearers, send representatives into the offspring of the next generation. Darwin's view, scarcely distinguishable from Lamarck's, was absolutely a statement for "the inheritance of acquired characteristics." Ultimately, however, Darwin equivocated on where these "sports," "mutants," or "heritable variants" came from. He simply did not know.

LATTER-DAY DARWINISTS
ON DAPHNE MAJOR

Darwin's intellectual legacy, spreading after his death from England, Germany, and the United States throughout much of the literate world, is still in vogue in myriad texts and classrooms as, in our opinion, the idiosyncratic belief system of most "modern evolutionists." It can be stated in abbreviated form as follows: All organisms derive from common ancestors by natural selection. Random mutations (heritable changes) appear in the genes, the DNA of organisms, and the best "mutants" (individuals bearing the mutations) in competition with the others, are naturally selected to survive and persist. The unsuited offspring die—they tend to be called "unfit" — with fitness, a technical term, referring to the relative numbers of offspring left by an individual to the next generation. The most fit, by definition, produce the largest number of offspring. The mutant variations then leave more offspring, and populations evolve; that is, they change through time. When the number of changes in the offspring accumulate to recognizable proportions, in geographically isolated populations, new species gradually emerge. When sufficient numbers of changes in offspring populations accumulate, higher (more inclusive) taxa gradually appear. Over geological periods of time new species and higher taxa (genera, families, orders, classes, phyla, and so on) are easily distinguished from their ancestors.

We agree that very few potential offspring ever survive to reproduce and that populations do change through time, and that therefore natural selection is of critical importance to the evolutionary process. But this Darwinian claim to explain all of evolution is a popular half-truth whose lack of explicative power is compensated for only by the religious ferocity of its rhetoric. Although random mutations influenced the course of evolution, their influence was mainly by loss, alteration, and refinement. One mutation confers resistance to malaria but also makes happy blood cells into the deficient oxygen carriers of sickle cell anemics. Another mutation converts a gorgeous newborn into a cystic fibrosis patient or a victim of early onset diabetes. One mutation causes a flighty red-eyed fruit fly to fail to take wing. Never, however, did that one mutation make a wing, a fruit, a woody stem, or a claw appear. Mutations, in summary, tend to induce sickness, death, or deficiencies. No evidence in the vast literature of heredity change shows unambiguous evidence that random mutation itself, even with geographical isolation of populations, leads to speciation. Then how do new species come into being? How do cauliflowers descend from tiny, wild Mediterranean cabbagelike plants, or pigs from wild boars?

Darwin's successors have returned to the rugged volcanic island of the Galapagos, off the coast of Ecuador, to watch evolution in action. Perhaps the most compelling real case of nonstop evolutionary change is the work of Professors Peter and Rosemary Grant on exactly those finches that were made famous by Darwin and his successors. The paucity of other birds on these remote outposts and the severity of environmental pressures led, somehow, to rampant and rapid speciation from common avian ancestors. As Jonathan Wiener (1999) describes in his book *The Beak of the Finch*, study of variation and the changes in these birds, isolated from the South American mainland for a million years, gives us the best traditional view of how the speciation process should work.

Today's ornithologists recognize thirteen species of Galapagos finches, which they place in four genera. The six ground-dwelling finches, who tend to only fly as a part of the mating game, spend their time hopping around. These ground dwellers are grouped into the genus *Geospiza*. *G. fortis* (strong) is the medium ground finch, *G. magnirostris* (big face) is the large ground finch. The sharp-beaked ground finch, the difficult one to distinguish, is called *G. difficilis*. *G. fulginosa*, the one who runs away, is called the small ground finch. The *large* cactus finch, *G. conirostris*, is probably so called because of its massive coneshaped bill suited to cracking large cactus seeds. The common name of *G. conirostris* is too close for comfort to the just-plain cactus finch *G. scandens*.

The other seven species divide into three groups: those who live in the trees on fruits and insects; strict vegetarians of the trees; and tree-dwellers who embody "convergent evolution"—they sing, act, and feed so much like warblers that they were at first taken to be warblers.

Since 1978, the Grants have camped on Genovese Island, by a sharkless, leechless lagoon named after Darwin. There they have measured songs, legs, eggs, and beaks of finches. All the islands of the Galapagos archipelago suffered the ultimate drought in 1977. Probably not a single drop of rain fell the entire year. None fell in the arid summer of 1985 either. Yet in the El Niño year of 1982–1983, the wettest year in living memory, there were some 200 millimeters of rain, with drastic effects on the birds and their food sources. Volcanic muds flowed and everything sprouted. Vines grew up tent poles, croton flowered seven times instead of once.

The number of seeds on the ground was more than ten times that of the year before. Caterpillars were five times as abundant and fatter than usual. The big cactuses, however, did not thrive. The heroes of the desert were inundated and overrun by small seeded plants. The birds copulated and bred and copulated and bred in never-before-seen orgies. This vast environmental change from dry to wet to dry again generated a numbers feast for the scientist measurers. Great population swings occurred—excursions away from the large cactus seedeaters and their starving young toward a baby boom among the small birds.

In nearly thirty years of work, the Grants have recorded some extraordinary changes in population structure and morphology. They have documented strong responses on the population and on the species level to the obvious selection pressures of wet and dry. They recorded changes in beak size along with alterations in the ability of these oral tools to crack open seeds. The differences in measured beak sizes between two highly selected groups of finches have gone from no difference (0 percent) to 6 percent. New work on birds by Mayr and Diamond (2001) shows unequivocal correlation of bird species with geographical isolation on Melanesian islands. Yet here's the rub: Speciation, the details of the appearance of *any* given *new* species of bird, whether Ecuadorian or Melanesian, has not been documented.

The differences in beak measurements between the six distinguishable species of ground-dwelling finches is about 15 percent. No changes of this magnitude, correlated with other traits that would produce a newly named species of Galapagos finch, were seen by the Grants—or anyone else. The Darwinian paradigm is operating exactly as it should: Different traits (whether within species or among different species) are varying in prevalence according to the demands of the environment. Obviously, the genes that produce these traits are varying in like fashion. But there is no evidence whatsoever that this process is leading to speciation.

Speciation, whether in the remote Galapagos, in the laboratory cages of the drosophilosophers, or in the crowded sediments of the paleontologists, still has never been directly traced. The closest science has come to observing and recording actual speciation in animals is the work of Theodosius Dobzhansky in *Drosophila paulistorium* fruit flies. But even here, only reproductive isolation, not a new species, appeared. The reproductive isolation occurred where a fully fertile population living at moderate temperatures became two populations—one cold-dwelling and the other warm-dwelling.

The best direct evidence for speciation, in our opinion, is the least known: It involves sexless beings and lies hidden in arcane professional literature. To see directly that a population of organisms speciates, one must look to the inhabitants of the microcosm. To study speciation we have to track symbiosis and the literature on symbiogenesis. The vast majority of those who write about evolution, the zoologists and other biologists, tend to be ignorant of the literature of symbiosis and of microbial and other community ecology.

The intrinsic limitation of darwinist literature was analyzed in 1999 by Douglas Caldwell, who began with Darwin's 1859 book itself. The terms used by Darwin and the number of times those terms appear include: "beat(s)"—seventeen; "death (dying)"—sixteen; "destroy (destroyed, destruction)"—seventy-seven; "exterminate (extermination)"—fifty-eight; "individual"—298; "kill (killed, killing)"—twenty-one; "perfect (perfection)"—274; "race (races)"—132; "select (selects, selection)"—540; "species"—1,803. By contrast, the following terms are absent from *The Origin of Species*: "association, affiliation, cooperate, cooperation, collaborate, collaboration, community, intervention, symbiosis." One hundred fifty years later, the habit of ignoring metabolic and physical associations between organisms persists. In a leading undergraduate textbook of evolution, *Evolutionary Analysis* (2001), by Freeman and Herron, the terms "combat," "competition," and "conflict" appear on at least eighteen pages, whereas "symbiosis" and "symbiogenesis" are not even mentioned once in over 700 pages. Erudite analyses of the "trees of life," many calculated by high-speed computer, bear only diverging branches. Few branches ever merge to represent symbiogenesis. The computer graphics may be impressive, but they do not reflect life's history on Earth, nor the evolution of new species.

HEREDITY AND HEALTH

For most of human history, physicians, shamans, and herbalists practiced the healing art in a state of well-educated ignorance. No science of genetics existed, even though astute practitioners knew that certain diseases prevailed in certain families. Among Hebrews in Biblical times, for instance, it was considered ill-advised to circumcise any boy whose uncle—but only on the mother's side—had suffered excessive bleeding at his own circumcision. To predict a woman's baldness or her baby's "cri de chat" (Tay-Sachs) disease, doctors traced the trait on both sides of the family. The great Hapsburg dynasty of Europe was identifiable for more than a hundred years by its members' jutting jaws. But practical genetic advice, culturally transmitted as learned folklore, never stemmed from a logical body of knowledge. Until the beginning of the twentieth century, nothing was known of the causes of inherited diseases—either their physiology or the reasons they ran in families—because no one, physician or not, knew anything about genes or the chemistry of DNA.

Conscientious medical men and midwives suffered enough plague, childbed fever, pox, and other "virulence" to firmly grasp the concept of contagion. But a coherent philosophy of contagious disease proved elusive. Men of medicine and court physicians knew the meaning of iatrogenic (from Greek for "doctor-caused") illness and made loyalty to their fellow practitioners a higher goal even than doing their patients no harm. Hospitals remained excellent final resting

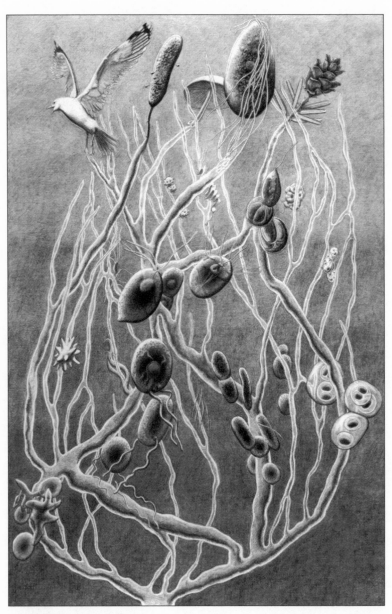

FIGURE 2.1   Tree of Life

places throughout most of European history. Neither health professionals nor the public knew of the existence of the microcosm, the world of the very small, until the work of Louis Pasteur and Robert Koch in the nineteenth century provided evidence of the identity of the culprits behind contagions. Only after Pasteur discovered the habits of bacteria and yeast and Koch developed his famous postulates for the proof that "germs cause disease" did the subvisible microbial presence begin to register within our cultural heritage. Only in the first half of the twentieth century, thanks in part to an influential book by Paul de Kruif, a journalist who worked at Rockefeller Foundation in New York City, did "microbe" become a term of common currency. Scientific curiosity is all to the good, but fear and greed are what move the masses. At least in the United States, not until de Kruif's *The Microbe Hunters* became one of the first popular science best-sellers and government pamphlets spread advice on antisepsis did young mothers and octogenarian physicians alike become utterly convinced that bacteria, or "germs," must be eradicated from our lives.

"Germs," like "weeds" or "toadstools," was an undefined but clearly understood term: It denoted carriers of disease, unwanted tiny forms of life. Expressed as "the enemy agent," the tenet became—as it is in the minds of many today—that all microbes must be vanquished. No difference between germ, virus, contagion, microbe, or bacterium was respected. The infant science "microbiology" still knew nothing of the distinct stages of the malarial parasite *(Plasmodium)* or of the nutritional interchange between *E. coli* and its human intestinal habitat. Physicians did little more than classify by symptom and prescribe safe treatment: Smoke, herbal infusion, aspirin, morphine, cocaine, extirpation, and many other "remedies" filled the black bag of tricks. Desperately ill patients were treated by "cupping," compresses, poultices, chanting, and laying on of hands. The major antidote, as from time immemorial, was confident reassurance dispensed at high prices by authoritative specialists, usually large men of commanding presence.

Evolutionary theory begins this twenty-first century in a condition much like that of medicine a hundred years ago. The search for new knowledge is inevitably embedded in culture. Professional evolutionary theorists tend to be abysmally ignorant of the three sciences—microbiology, paleontology, and symbiosis—most relevant to their work. But the situation in evolution is even worse than it was in nineteenth-century medicine. The difference is that whereas the facts required by doctors at the turn of the eighteenth to nineteenth century did not yet exist, the background essentials for evolutionists at the turn of the twentieth to twenty-first century exist but are systematically ignored.

Science has documented evolution in action even if most scientists are not aware of the fact. Most information relevant to understanding evolution lies hidden in arcane literature. The news does not reach either the professionals or the public. A fragmented body of literature, detailed but disorganized, does document the ways in which species originate and case studies exist that follow speciation. The marvelous details of how one species leads to another, or vanishes permanently, need spokespeople. How does complex novelty in living beings first appear? How does it spread? How do plant, animal, and microbial bodies change over time? This story has not been written because most of the notes for it are recorded in the esoteric languages of biochemistry, microbiology, and other sciences. The species-origin tale, and its relation to isolated populations as first explained by Mayr (1942, 1982), has been mostly inaccessible to the reading public as well as to natural historians and science museum designers. Even among biologists and geologists who write about evolution there is a paucity of documentation and explanation of case histories of the origins of any given species. Botanists, students of the plant world, in occasional cases can correlate chromosome and chloroplast differences inside plant cells with the appearance of new horticultural or agricultural species. But since they do not deal broadly with evolution nor do they generalize their explanation to animals, algae, or bacteria their descriptions tend to be limited to their progressively shrinking world of "plant sciences." Geneticists, ecologists, microbiologists, physiologists and other laboratory dwellers, and experimentalists tend to avoid discussion of the evolutionary implications of their work. Most of them simply have no idea how the complexity of life evolved, or in any case they do not write about it. For some, the cultural heritage they absorbed with their graduate training disdained evolutionary history as "speculation." Many of these "hard-nosed" practitioners do not even consider evolutionary studies a science in the league with "hard sciences" such as molecular biology or biochemistry. Nearly all scientists, perhaps even all, prefer to tackle questions answerable by direct evidence. Reconstruction of the history of life, the major activity of the evolutionist, is often dismissed as unprovable speculation. Philosophy and history are viewed as questionable traditions, useless or leisurely occupations, in modern scientific circles, especially in the United States. Active scientists often deny they even have any philosophy—they say it interferes with their work. The scientific reconstruction of prehistory, however, an intrinsically philosophical practice replete with inference and judgment, is essential to the answer of the central question of how species began.

We are also impeded by the isolation of scientists. Even the communication between those fields relevant to evolution, such as paleontology and molecular biology, is exacerbated by terminological differences. All scientists, no matter how gifted, can only concentrate closely on one thing at a time. The day-to-day scientist lacks any way to integrate into his or her rigorously bounded studies the myriad idiosyncratic descriptions of fundamental discoveries in live organisms. We hope here to lay some communication lines with as little distortion and over-simplification as possible.

The phenomenon of evolution occurs over the entire face of the Earth, from at least fourteen kilometers below sea level, in the ocean's deepest abysses, to as many as eight kilometers above sea level, on the high mountains. Bacterial life has been recorded in wet fissures of granitic rocks at depths of at least three kilometers into the Earth's lithosphere and at hot water vents on the sea bottom. The biosphere, the place where life exists, then has overall dimensions of a hollow sphere some twenty-three kilometers wide. This, compared to the 6371 km radius, corresponds to only 0.0007% of the volume of the planet.

Life began probably more than 3,800 million years ago. The detailed record of evolution, preserved in rocks both as fossils and as short- and long-chain extractable carbon compounds, overwhelms those who study it. Cellular life reflects its evolutionary history. Yet in spite of the diversity of clues by which the evolution of life is reconstructable, most self-described evolutionary biologists disregard cell biology, microbiology, and even the geological rock record. Many are so preoccupied with land-dwelling animals that they continue to believe that no record of evolution exists prior to that of the last 541 million years. The crucial science of the Archean and Paleozoic eons, the worldwide geographies and details of metabolism, and cell chemistry hold the answer to Darwin's enigma, but these vast fields of inquiry are formally eliminated from consideration.

As biochemistry, microbiology, and electron microscopy have been factored in, evolutionary concepts have necessarily become far more multidimensional and interconnected. The immense expansion of the literature of molecular biology, especially the sequence data of proteins and nucleic acids, molecules that are present always in all live beings, has brought huge new insights to Darwin's vision. The details of molecular evolution support the generalization that the amino acid sequences in proteins and the nucleotide sequences in DNA and RNA are related by common ancestry. But the abstract dominant gene "A" that mutates to its recessive alternative (allele) "a" as neodarwinist theory purports is a vast, even useless overgeneralization for real genetic change in populations. Rather than idealized formalities of darwinian "modern synthesis," the organizing principles for understanding life require new knowledge of chem-

istry and metabolism. Insights into the working of cells have clari-
fied the modes of evolution since Darwin and his immediate followers wrote their analyses. The results of new laboratory and field
science contradict, bypass, or marginalize the formalism of Neodarwinism except for variations within populations of mammals and
other sexually reproducing organisms. The mammals probably constitute only one ten-millionth of all species living today.

From different vantages this poorly known idea can be gleaned:
The agents of evolutionary change tend to be fully alive organisms,
microbes, and their ecological relations, not just the random mutations these microbes have inside them. The formation and diversification of any new species is the outward manifestation of the actions of subvisible forms of life: the smallest microbes, bacteria,
their larger descendants, the larger microbes, protists,* and fungi,
along with their intracellular legacies, organelles such as mitochondria and centrioles. Evolution emerges from the fact that these small
living organisms and their progeny tend to outgrow their bounds.
The unseen beings that decimate our populations with virulent disease and provide soil nitrogen to our food plants play the major creative role in the genesis of new species.

CELLS, MICROBES, PROPAGULES, AND AUTOPOIETIC UNITS

The minimal unit of life is the cell. All microbes are composed of
cells. They can all be unambiguously classified either as bacteria
(without nuclei) or protists or fungi (both of these groups are composed of cells with nuclei). The smallest and least complex cells,
those of some bacteria such as *Mycoplasm geniticulum,* contain approximately 500 genes.

Some people, indeed some authors, present viruses as the smallest forms of life. But viruses are not alive and indeed they are even, in
principle, too small to be units of life. They lack the means of

---

* Protists are the smaller members of the kingdom Protoctista see pages 142–143.

producing their own genes and proteins. Viruses behave as chemicals until they enter the appropriate living cells, where they can co-opt the amenable cellular environment to reproduce themselves. On its own a virus is inert. Cells, however, when they have access to material and energy in usable form and are bathed in suitable water, grow—usually to twice their size—and reproduce, usually by division. Cells display metabolism, the interconnected chemical reactions that maintain their processes. Genes by themselves, like viruses, are unable to produce cell material, which is mostly protein. Genes and viruses absolutely require the intact live cell for growth, reproduction, and propagation. Whereas the smallest cells may have fewer than 500 genes (and therefore make fewer than 500 different proteins), most bacterial cells contain between 2,000 and 5,000 genes. Some larger, more complex oxygen-respiring bacterial cells have nearly 10,000 genes. These bacteria therefore approach the numbers of genes typically found in the cells of nucleated organisms. Nucleated cells, the minimal live components of all large organisms (fungi, animals, plants and their protoctist ancestors) usually contain more than 10,000 genes but fewer than 100,000. Human cells may have 30,000 to 60,000 genes. Those of the yeast we use for beer and bread making, *Saccharomyces cerevisieae,* have only 8,000.

"Autopoiesis," literally "self-making," refers to the self-maintaining chemistry of living cells. No material object less complex than a cell can sustain itself and its own boundaries with an identity that distinguishes it from the rest of nature. Live autopoietic entities actively maintain their form and often change their form (they "develop"), but always through the flow of material and energy. For any organism, any autopoietic entity, a specific sustaining source of energy (such as visible light, methane oxidation, or sulfide oxidation) can be identified, as well as a source of carbon (such as sugar, protein, carbon dioxide), nitrogen, and other required chemical elements.

At temperatures near absolute zero, when movement of molecules stops, metabolism, the manifestation of autopoeisis, also ceases.

However, the material structure of many live cells is such that they retain the capacity to resume metabolism when the temperature is raised again. Bacterial spores and carefully frozen embryos do not lose their intrinsic organization when frozen, and upon warming they are fully capable of the metabolic transformations that underlie self-maintenance. Hence we can see that although life fully blooms only as a process of flowing matter and energy, it can remain dormant in the organization of living matter.

Microbes are champions at passing their DNA to others in the form of entire functional genes. These machinations underlie the story of darwinian evolution. Microbes living on their own, under conditions of stress and deprivation, tend to merge with other forms of life. Some of these associations last for a season or less, but occasionally microbes and larger life forms fuse permanently. Lamarck was correct: Acquired traits can be inherited not as traits but as genomes. But Lamarck, and Darwin with his pangenesis idea, were both wrong when they suggested acquisition might be the fate of any characteristic. The only "character" or trait that can be passed down (vertical inheritance), or acquired (horizontal inheritance) and then propagated from generation to generation, is a "character" encoded in genes. This means that "characters" capable of propagation in the environment, of acquisition and inheritance from one generation to another, tend to be complete genes in genomes. Genomes are entire genetic systems, active only when they reside inside cells. Alien genomes can be, and often are, inherited by others. Live beings composed of cells can and do pass on their sets of genes to others. Why have human genetic engineers been so successful in passing foreign genes to food plants and domestic animals? Because nature has indulged in these gene-trading and genome-swapping tricks for eons. The real genetic engineer is the microbe; the scientists and technicians are merely the go-betweens.

That new species form by inheritance of acquired microbes is best documented in protists. We know that the process of inheriting a foreign set of genes is not streamlined, nor is it planned in advance. Nature tends to be opportunistic, not foresighted. But genome acquisition in many cases has left so much evidence behind that its circuitous routes can be retraced. Most genetic takeovers and acquisitions, mergers and fusions ensue under conditions of environmental hardship. The complexity and responsiveness of life, especially the appearance of new species from mergers between and among differing ancestors, can be understood only in the light of a peculiar evolutionary history. For every case the questions must be asked: When, where, and in what populations of organisms did the new species evolve? How often, with what frequency, and in what groups of organisms is genome-acquisition the major mode of speciation? These are research questions we hope to inspire.

Today's evolutionary saga, legitimately vulnerable to criticism from within and beyond science, becomes incomprehensible if the microbes are omitted from the story. Just as chemistry would be incomprehensible with no mention of atoms, evolutionary talk is largely meaningless without an understanding of the microcosm.

THERMODYNAMICS, "PURPOSE," AND EVOLUTIONARY "PROGRESS"

Acceleration of change sweeps us away. One week recently one of the authors (Lynn Margulis) passed a yellowed, run-down, student-infested house on the main street of town while bicycling to the campus, as usual. A week afterward the view past the house across to the neighboring schoolyard was splendid. For the first time in living memory, it was unobstructed. Mountains framed the distant backdrop. The house was gone. No sign that any house had ever been there remained, except newly raked soil in the footprint of a simple ground plan. Such dramatic changes in our immediate surroundings are commonplace. Burger King, Toys "R" Us, Wendy's, McDonalds, and bank branches sprout in our cities and towns. Mom-and-pop shops, wheat fields, and old oaks disappear like coins dropped into the sand. Native Americans thrive if they form gambling liaisons, graduate students receive stipends when they change from studying the habits of beavers inside their lodges to the search for genes in the human genome. Why do the forces of change always seem to prevail over the quiet and uneventful habits of the past? Why does the evolution of life seem to accelerate as we move into the present and out toward the future? Is evolution just random change? Does not the evolutionary process itself, the origin and diversification of life from common ancestors, seem to be directed? When we ask evolutionary biologists and other scientists if the evolution of life is going in some direction, they adamantly deny it. But our everyday experience suggests that our social environment grows more complex. Our natural green and watery environment seems to shrink to be locally augmented with metallic solids: Neon lights, traffic signals, and other aspects of urbanization replace woodlands and open streams at an ever-increasing rate of change. People crowd out foxes and antelope, pigeons and sparrows replace orioles and woodpeckers. Digital tools supplant simple mechanical devices at alarming speeds. Evolution of life *does* seem to have a direction. Life's peculiarities and human technologies do seem to expand at an accelerating rate of change as we come from the past toward the present.

A branch of mathematical physics called thermodynamics may be useful to us here. Thermodynamicists describe machines or chemicals in closed boxes. Traditionally, they have worked with energy, temperature, and heat and, as physical scientists, have avoided wet and wiggly beings. But we are in the throes of a great breakthrough: the linkage of the exact science of thermodynamics with other sciences such as biology, meteorology, and climate change. Nature takes her course and scientists learn, after many failures and false starts, to accurately describe nature in mathematical and chemical terms. The new thermodynamics of life is likely to have a lasting influence on evolutionary thought. It connects living with nonliving matter in a comprehensive way.

We all know that "nature abhors a vacuum," and therefore air rushes in to fill any void we make with a pump. Cans deprived of air will collapse immediately. We know too that gravity attracts matter toward the center of the Earth and that therefore bodies released from heights fall down. We take the predictable fall of matter to ground as evidence of a law of nature.

The new thermodynamics embraces both these ideas in the phrase "nature abhors a gradient." As we learn to understand this fundamental new idea a whole new perspective is let loose: Life behaves according to laws of nature in exactly the same way as air molecules rushing to fill the vacuum or falling bodies coming to Earth. So what is this new thermodynamics and how does it work?

In *The Way of the Cell* (2001), Franklin Harold wrote, "Those who envisage a fundamental link between the thermodynamic arrow of energy dissipation and the biological arrow of the greening earth make up a small minority. . . . But if their vision is true, it reveals that deep continuity between physics and biology, the ultimate wellspring of life." We happen to be among those who think that this vision is true. Evolution is a science of connection, and connection does not stop with ties of humans to apes, apes to other animals, or animals to microbes: Life and nonlife are also connected in fundamental ways. We have seen that the organization of life is material. But energy also organizes.

Eric D. Schneider, a Columbia University geologist who went on to serve in the U.S. Environmental Protection Agency in the 1970s, is one of the chief visionaries of the new thermodynamics. He became disappointed with the primitive level of environmental science at the E.P.A. when he saw that hardy fish, placed in aerated labeled pickle jars, were fed toxins to set standards for what bureaucrats determined as "appropriate pollution levels." Schneider, who thought environmental protection should be based on adequate descriptions of ecological phenomena, devoted decades to research on how nonlife connects to life. Drawing on the works of such scientists as L. Boltzmann, V. I. Vernadsky, I. Prigogine, J. E. Lovelock, and H. Morowitz, his findings are simple and revelatory: Life is one of a class of systems that organize in response to a gradient. A gradient is defined as a difference across a distance.

All of these systems produce unexpectedly complex features. Temperature gradients produce highly organized fluids, hexagonal convection currents called Bénard cells. Air pressure gradients cause spinning tornadoes and global weather patterns that disappear once the barometric pressures are equalized. Chemical gradients create complex chemical "clocks." A crucial point here is that not only is life consistent with the second law of thermodynamics—that entropy increases in isolated systems—but that complex systems (from tornadoes that equalize pressure gradients to organisms that reduce solar and chemical energy gradients) reduce the gradients around them more effectively and quickly than would be the case were they not to exist. An atmospheric pressure gradient, for example, will take longer to reach the state of randomized chaos without the complex cycling system of a tornado whose "function" is to achieve this natural end.

The original gradients described by science were in steam engines. Differences between hot and cold (which people had long noticed tended to equalize on their own accord to become lukewarm) could be usefully converted into energy by machines. The mathematics of how to convert heat to work became the basis of the science of thermodynamics. The term comes from "heat" and "motion" in Greek. Thermodynamics is the study of how systems handle and transform energy. Classical thermodynamics studied isolated and more-or-less closed systems, systems that ultimately lost their function and whose component molecules became randomly arrayed to the point where they could no longer do work.

But living beings are not steam engines or closed mechanical systems. We are open systems, organized by the energy and materials that incessantly flow through us. Thus we (life in the extended genetic sense) are not destined to revert to the random state measured either in isolated chemical systems or machines left on their own. The difference between open, living systems and the artificially closed systems of classical thermodynamics once made this science of energy seem irrelevant to life. Worse yet, Darwin's theory of evolution by natural selection was naively thought by too many in the nineteenth and twentieth century to mean that the laws of thermodynamics contradict the phenomenon of life. Thermodynamic systems become random, disorganized, less complex over time, whereas living systems apparently increased in levels of organization and complexity. Now, the work of the new thermodynamicists has resolved the apparent contradiction between the tendency to disorganization of inert matter and the tendency to organization of living matter. Life does not exist in a vacuum but dwells in the very real difference between 5800 Kelvin incoming solar radiation and 2.7 Kelvin temperatures of outer space. It is this gradient upon which life's complexity feeds. Like the thermal Bénard cells, which stabilize by defying statistical tendencies to approach equilibrium because they manifest the improbable gradient around them, or the meteorological whirlpools that maintain their structure "to" get rid of an atmospheric pressure difference, life too is a gradient-reducing system. These thermodynamic ideas connect life to nonlife, just as Darwin's natural selection connects us to apes and eventually to microbes. The power of thermodynamics to explain the overall features of life's behavior is not trivial; it is part of the future of science.

Thus life does not contradict the well-known second law of thermodynamics. No new law of thermodynamics is needed once we realize that classical thermodynamics' study of complex behavior used the artificially limiting case of closed and isolated systems. When classical thermodynamics is extended to an open cosmos, complexification becomes comprehensible. Rather than bumpily coining a special new "fourth law" to apply to life, as Stuart Kauffman proposed in his 1999 book *Investigations*, we can smoothly apply the general thermodynamic view to all matter. Technology, whether human, bower bird, or nitrogen-fixing bacterium, becomes the extension of the second law to open systems. The second law, describing an inexorable increase in a measure known as entropy—first formulated as "heat divided by temperature" and later reworked into the statistical tendency of more likely arrangements of matter and energy to occur with the lapse of time—does not contradict life's local tendency to increase order. No. Exactly the opposite is true. Creationists are mistaken. Life's origin, its cellular organization, its expansion, and evolution of increasing biodiversity are entirely consistent with the new thermodynamic thought. The remarkable but not miraculous arrangements of living matter provide another example, along with cycling hurricanes and chemical clocks, of how nature builds structures to reduce gradients. Most evolutionists argue that life originated on Earth—perhaps at a chemical gradient between hydrogen-rich compounds and other carbon and oxygen-rich substances. These are combinations that burn. Or perhaps life started in the light, where the sun's high-energy photons could be cycled through the precursors of metabolism before modern metabolism fully evolved. Life developed the identity of the cell. Living matter in the form of cells and organisms made of cells began to reduce chemical and energy gradients on a turbulent early Earth. Indeed, the tendency to locally organize "in order to" get rid of the statistical anomalies incarnated by gradients is profound and organizes life at all scales. The superior efficiency of complex ecosystems at reducing gradients is measurable and has been measured, for example, by airborne thermometers that show the superior ability of tropical forests relative to grasslands and deserts to cool themselves, thereby reducing the solar electromagnetic gradient. This is no vague and abstract theory of complexity, but a tested hypothesis: As measured both by low-flying airplanes and by satellites, ecosystems are cooler when they are more mature and biodiverse. Ecosystems begin with fast-growing colonizing species and at first are relatively inefficient gradient reducers. But as they mature, the energy and material cycles become larger in scope. New organisms enter the system and establish new habitats. Growth slows down, but overall the integrated ecosystem is better than its predecessor at reducing the gradient between the sun and space. The fact that mature tropical ecosystems stay cool displays the system's power of gradient reduction.

Not only the development of ecosystems but the entire direction of evolution is informed by the thermodynamic mandate, the "law of nature" to reduce ambient gradients. In ecosystems, evolution, and economies we note an increase in complexity in, for example, the total number of species spread over a larger volume, increasing differentiation of "division of labor," more developed and complex computer networks, and increasing functional integration of material and energy flows such as gas, electricity, water, and sewage in the urban environment. The increase in the abilities of organisms to adjust themselves to dwindling resources as we come to the present is a further example. Although it is difficult to say why the universe is so organized, the measured universal expansion since the Big Bang of space continues to provide a "sink" (a place) into which stars as sources can radiate: A progenitive cosmic gradient, the source of the other gradients, is thus formed by cosmic expansion. For the foreseeable future the geometry of the universe's expansion continues to create possibilities for functionally creative gradient destruction, for example, into space and in the electromagnetic gradients of stars. Once we grasp this organization, however, life appears not as miraculous but rather as another cycling system, with a long history, whose existence is explained by its greater efficiency at reducing gradients than the nonliving complex systems it supplemented.

This new thermodynamics (sometimes called homeodynamics) lets us begin to glimpse the path from matter (gradient breakdown) to mind (gradient perception)—from energetic to informational "self"-organization. We put "self" in quotes here because, in fact, complex systems are "other"-organized, precisely *not* "self"-organized. The tendency of systems to organize comes from the gradients in their immediate surroundings, not from their own internal components. The informational structure of life, carried in DNA, has become self-sustaining via reproduction. But memory exists as well in fully nonliving systems such as vortices organized by rotational pressure gradients. These jump to new states and new values that depend on their history. Thus it seems to us that, without invoking any vitalism or mysticism or spiritualism, we can recognize in ourselves a "purpose." This purposefulness is an offshoot of the thermodynamic tendency to come to equilibrium. Complex systems, life included, tend to arise in order to bring their gradient-rich surroundings to equilibrium.

In the eyes of many Christians, the darwinian revolution left nature purposeless, at least on paper. Darwinians, faced with a personal Creator as the only conceivable source of purpose, hastened to agree. But physical purpose is more subtle than that. From the thermodynamic vantage point, purpose has a physical aspect. It is no more uniform than memory, which manifests itself in bodies, genetically, and brains, neuronally—and even in machines, magnetically. And like memory, purpose—with its orientation toward the future—has a thermodynamic genesis.

Life is thermodynamic. A continuous whirlpool downstream of Niagara Falls has a name: "Whirlpool." We give names to things, like species and hurricanes, that keep their identities—at least for a time. The formation of stable identities aids the thermodynamic process of gradient reduction. The highly heritable members of a species, like other cyclical and complex thermodynamic agents, provide stable vehicles of degradation. The cycling selves of life survive in order to reduce the energetic and material gradients that keep them going; they covet and tap into these gradients to survive long enough to reproduce. As natural selection filters out the many to preserve the remaining few, those few ever more efficiently use environmental energy to "purposefully" reduce their gradients. The key point is that living and nonliving "selves" come into being to reduce gradients naturally. The reproducing self of biology is a higher-order cycle whose antecedents can be inferred from the cycles of the nonliving world. Nucleotide replication and cell reproduction do not emerge from nowhere. They are born in an energetic universe from thermodynamic tendencies inherent in nature.

CHAPTER 3

RELATIVE
INDIVIDUALITY

We recognize individuals with ease. A group of individuals of the same species, in the same place at the same time, we call a population. Taken together with other forms of life, for example food plants and animals, different populations at the same time and place are recognized to form communities. Sometimes the largest and most dominant members of communities may be smaller than a millimeter in their largest dimension; if so, we speak of microbial communities. No life on Earth consists of unassociated individuals of the same population, like jailed adolescents or rows of corn plants—at least not for long.

INDIVIDUALITY FROM COMMUNITY

Communities are natural groups; in the natural world we recognize them easily without formal training—treetop communities, pond

water communities, shoreline communities, cliff-dwelling or woodland communities. A rule of thumb that we all use unselfconsciously is that individuals who seem to have everything important
in common, individuals who seem "the same" in all major aspects of
their lives, belong to the same species.

In the eighteenth century, Linnaeus and other, primarily European naturalists began to name and document species the world
over. A flurry of documentation and publication began as Euro-
peans explored the tropics of America, Africa, and Southeast Asia.
The Linnaean task of safe documentation and proper naming of all
organisms has never been completed. Even now an effort is under
way to encourage the monied interests of global corporations to use
their profits for an "All-Species Inventory." With the international
use of high-speed computers, satellite technologies, and real-time
transfer of copious quantities of data, the leader of this project, naturalist Peter Warshall, has argued that we ought to record the
world's living diversity before we destroy it. Extinction of species,
like a death in the family, is an irreversible loss.

Even while the "All-species whole-Earth data set" is far from
complete, some salient features emerge. Some ten to thirty million
different species of organisms are estimated to be alive today,
whereas fewer than two million of these have been documented in
professional literature. Species of extinct organisms, registered in
the paleontological literature based on fossil evidence of their former existence, number only about 150,000. Most scientists concur
that over 99 percent of species that ever lived are extinct. Hence the
guess for the number of species that have ever come or gone, since
life on the Earth began 4,000 million years ago, is 200,000,000
million. No one we know can distinguish more than a few thousand species, and in fact we know almost no one who can even recognize that many, even with the aid of appropriate library books. In
the field, especially in the flagrantly diverse new world tropics, most
of us are hopeless. Even Mayer Rodriguez, consummate Ecuadorian

guide on the Tiputini branch of the Napo tributary of the Amazon river, cannot identify more than about 500 species. And no one knows that region better than he does.

All known organisms can be placed unambiguously into one of two inclusive groups that depend upon the types of cells of which they are composed. The first, presumably earlier group is by far the more diverse (from the viewpoint of metabolic modes) and essential to the environment at the planet's surface. These are the bacteria, all of which are composed of cells that lack nuclei (prokaryotic cells). The familiar life forms (animals and plants) as well as the two groups of smaller but visible beings (fungi and protoctists) belong to the second, newer group. These larger life forms are known as eukaryotes, truly nucleated organisms, because the cells of their bodies contain nuclei.

Hence life on Earth is neatly classifiable into five groups that we, in the great tradition of biological taxonomy, consider to be kingdoms. (Table 3.1) Kingdoms are the "highest," which only means the most inclusive, taxa. This tradition uses all information available about an organism to place it in a group as reflective as possible of its evolutionary history. Very briefly, members of the bacteria kingdom are composed of small cells with threads of genes, not bound to protein, called genophores or chromonemes. When seen in electron microscopes the genetic structures are called nucleoids, to distinguish them from the nuclei of eukaryotes. The other four great kingdoms of life (the nucleated organisms whose DNA, wrapped in protein, is inside membrane-bounded nuclei and packaged into chromosomes) can be summarized too. First we have the embryo-forming groups: plants, which grow at one stage in the tissues of their mothers and at another stage from spores that contain only a single set of chromosomes; and animals, which grow after fusion of an egg with a sperm to form an embryo called a blastula. The less-known eukaryote kingdoms are the fungi (molds, mushrooms, yeasts, and their relatives) and the protoctista, unruly microbes (protists) and their larger descendants that gave rise to the

Table 3.1—Kingdoms: Largest Groups of Living Organisms

Time*	Type of cell	Chromosome sets	Comments	Examples
BACTERIA (3500)	Prokaryotic (no nucleus)	None. Chromo-nemal genetic organization	unidirectional DNA transfer from donor to host	*E. coli* "bluegreen algae" = cyano-bacteria; archaebacteria
PROTOCTISTS (2000)	Eukaryotic (membrane-bounded nucleus)	haploid (1) and diploid (2), variable	variations abound, cell and nuclear fusions	algae, amebas, ciliates, slime molds
FUNGI (450)	Eukaryotic (membrane-bounded nucleus)	haploid (1)	grow from zygo-asco- or basidio-spores, chitin cell walls	molds, yeasts, mushrooms
PLANTS (450)	Eukaryotic (membrane-bounded nucleus)	haploid (1) and diploid (2)	maternally retained embryos, cellulosic cell walls	mosses, ferns, flowering plants
ANIMALS (600)	Eukaryotic (membrane-bounded nucleus)	diploid (2)	blastula embryos, no cell walls	mollusks, arthropods, fish, mammals

*Approximate time of appearance in the fossil record, measured in millions of years before present

fungi, animals, and plants. These latter organisms form some fifty natural groups, of which most people have only heard of three or four. Among them are the red and green seaweeds, the brown algae including the giant kelp, the slime molds and water molds, shelled foraminifera, glassy diatoms, ciliates like *Paramecium*, the amebas, and the euglenids.

For the numbers of living species of these kingdoms we have only crude estimates that may be wildly incorrect. Animals, proba-bly because people are good at distinguishing beetles, dominate. Over 10 million—perhaps as many as 30 million—are thought to exist. Some 500,000 plants, 100,000 fungi, and 250,000 protoctists are suggested to be lurking in the woods and waters of this world. As for bacteria, although thousands have been named as species and no doubt thousands can be distinguished, the species concept doesn't apply. Although bacteria can be grouped on the basis of common features, these groups change so quickly that they are never fixed and recognizable like eukaryote species. Bacteria pass genes back and forth. All can simply reproduce, and thus at any given time have but a single parent. The intervention of sex, the formation of a new bacterium with genes from more than a single source, is a unidirectional affair. The genes pass from a donor individual to a recipient . . . but donors can change to recipients and vice versa in minutes. Furthermore the gene swapping is entirely optional. If a bacterium can survive and grow under conditions in which it finds itself, sex is dispensable at all times. Indeed bacteria are willing and able to "have sex" with naked DNA molecules that they absorb from the water in which they are bathed.

Life originated with bacteria; therefore we can say that the origin of life was concurrent with the origin of bacteria. But we agree with Professor Sorin Sonea and his colleague Lucien Mathieu, of the Université de Montreal, that bacteria do not have species at all (or, which amounts to the same thing, all of them together constitute one single cosmopolitan species). Speciation is a property only of nucleated organisms. It began with the earliest protoctists, long after bacteria had evolved nearly all the important metabolic traits displayed by life on Earth. Thus the origin of species itself was not at all concurrent with the origin of life; rather it occurred long after, in the Proterozoic Eon. Individuality does not come exclusively from diversification and branching evolution, as the neodarwinists would have us believe. It comes equally frequently, if not even more often, from the integration and differentiation of fused beings, once independent, but over time individualized and selected as wholes. This simply is another way of stating our thesis: speciation by (and in the aftermath of) symbiogenesis.

The creative force of symbiosis produced eukaryotic cells from bacteria. Hence all larger organisms—protoctists, fungi, animals, and plants—originated symbiogenetically. But creation of novelty by symbiosis did not end with the evolution of the earliest nucleated cells. Symbiosis still is everywhere.

Many examples of evolution by symbiosis strike us as remarkably beautiful. Pacific coral reefs such as the Great Barrier of Australia represent associations between modern (scleractinian) coral and dinomastigotes such as *Gymnodinium microadriaticum*. New England lichens, New Guinea ant plants, and even milk cows serve as examples of the power of living fusions. Members of different species, and in the case of cows and corals, even of different kingdoms, under identifiable stresses formed tightly knit communities that became individuals by merger. Details abound that support the concept that all visible organisms, plants, animals, and fungi evolved by "body fusion." Fusion at the microscopic level led to genetic integration and formation of ever-more complex individuals. The thermodynamic drive toward more complex gradient-reducing systems finds expression in the continual creation of newer, more intricate forms of association between life forms, including symbioses.

STRENGTH IN NUMBERS

Terry Erwin, a professor at the University of Alabama who shakes insects from nets in the Amazon canopy and counts them, tells us that, routinely, two-thirds of the species he finds are new to science. If ten million species of animals are so far documented in the annals of the learned, and if his counts represent the world, he suggests that at least thirty million living animal species must exist. Joseph Leidy, sage of nineteenth-century Philadelphia and one of the founders of that city's Academy of Natural Science, summarized his philosophy of a good life: "How," he wrote, "can life be tiresome when there is still another rhizopod to describe?" Whether counted in Leidy's rhizopods (amebas), Erwin's beetles, or the thousands of orchid species of the mountains of Colombia, life's diversity staggers the imagination. The Linnaean task of classifying all living things is as unfinished in the twenty-first century as in the eighteenth when Mrs. Linnaeus sold her husband's whole collection to London to pay off his debtors and his daughter's dowries. His jars, bottles, and dry plant specimens still reside in the basement of the Linnaean Society in Burlington House, just off noisy Picadilly Circus.

The goal to catalogue species of life on the planet remains noble, indeed more noble and useful than in Linnaeus's times when species were catalogued and classified in a practical way without regard to preservation of planetary biodiversity. He never believed the forms of life reflected an evolutionary history but that all biological abundance and diversity were the works of a good and prodigious deity. His concept of species was of a fixed and unchanging kind, identifiable by visual characteristics. Most of the names he gave to his 10,000 species are in current use. *Panthera leo* is the name he gave to lions and *Felix catus* is what he called our cats. *Malus deliciosus* was how he recorded our apples and *Mytilis edulis* referred to the delicious blue mussels of the Atlantic shore. The first appellation, always capitalized, refers to the larger group or "kind" of life (people, dogs, apples, mussels)—the genus—and the second, the "specific epithet," indicates the species (wise people, familiar dogs, delicious apples, and edible mussels). As for fungi *Lactarius deliciosus* (delicious milky mushroom) or *Penicillium chrysogenum* (little golden pencil), these of course, since they weren't animals, were for Linnaeus plants.

Species, in short, were and still are the lowest common denominator. From the days of Linnaeus on, species have been "banked" in the literature: the name, the place of collection, the published description, and its author are publicly deposited in an herbarium or a natural history museum. As familiarity with microscopic life abounded, species proliferated and naming practices grew to include bacteria (as plants, even when they swam) and other unicells as plants (if they were green or greenish) and animals (if they swam).

Many millions of words have been written on the definition of species. Here we suggest a new testable idea about what species are.

SPECIES DECONSTRUCTED: MIEHE'S ARDISIA, AND ROSATI'S EUPLOTIDIUM

The existence and extinction of species and the origin of new ones are phenomena found only in nucleated organisms. This new concept is a simple corollary of Sonea's idea that prokaryote populations in nature do not form species. Prokaryotes, whether archaebacteria or eubacteria, by contrast all belong to a single worldwide species. The bacterial Internet preceded ours by 3,000 million years! Organisms with bacterial cell organization are not classifiable into distinct species. Nucleated organisms, all of which are products of symbiogenesis, first appeared on Earth more than 1,200 million years ago. Their appearance correlates entirely with the appearance of the first species and genera. Why? In our opinion, because:

An organism (A) belongs to the same species as another organism (B) if and only if A and B have precisely the same cellular ancestors, that is, they are descended from the same genomes, and the relations between these genomes are the same. In other words if we make a list of the genomes that constitute A and of those for B, the lists will be the same if organisms A and B are assigned to the same species. The lists will differ if they belong to different species.

Zoologists have long recognized the validity of Ernst Mayr's biological species concept: Two organisms belong to the same species if in nature they recognize each other, mate, and produce fertile offspring. This definition, which is apt for most animals and many plants, is one with which we concur. We see it as a special case of our definition. Only those animals that share the same complement of genomes can develop complementary genders that can indulge in

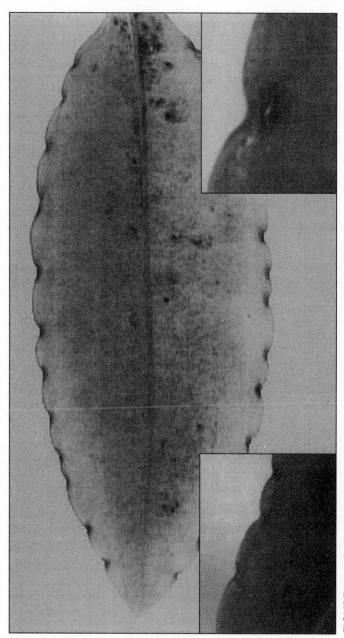

FIGURE 3.1   *Ardisia* Leaf Margin Fluting Due to Bacterial Colonies

fertile matings. Our definition, however, recognizes the heteroge-nomic composition of all nucleated organisms and is thus far broader. Our "species-component list" approach acknowledges the existence of tens of thousands of species of protoctists, fungi, plants, and animals that don't mate to produce offspring.

Throughout this book we review examples of new species and higher taxa that originate by acquisition and incorporation of for-merly independent genomes. Here are two examples of how one may analyze species from their component genomes.

Miehe's Leaf Margin Secret Confirmed

New species and genera of flowering plants evolved when the leaves of these plants acquired and integrated into their life histories a bac-terial genome. In the family of flowering plants called the Myrsi-naceae are three genera found in Indochina, Formosa, Malacca, the Malay peninsula, and the Philippines. The members of the largest genus, *Ardisia*, have fluted leaves (Figure 3.1).

Ardisia's secret has been known since von Hohnel's work of 1882 and has been confirmed by several first-rate investigators, no-tably Hugo Miehe from the botanical garden at Buitenzorg in 1910 and his successor Ph. DeJong from the botanical institute in Leyden in 1938. The fluted-leaf morphology has a bacterial basis. All of the thirty species constituting the genus *Ardisia*, plus those of its rela-tives *Amblyanthopsis* and *Amblyanthus,* have specific bacteria living in their leaves. These bacteria can be isolated and grown in culture. They are heterotrophic symbionts and are inherited through the flower and the plant embryo in predictable fashion. They are not intracellular—that is, they are never inside the plant cell—but are found between the cells in all the leaf nodules. In the seeds the bac-teria live between the embryo and the endosperm, the storage mate-rial encasing the embryo. When deprived of its bacteria by, for ex-ample, a heat treatment that kills the bacteria but not the plant,

Ardisia grows as a cripple. The treated plant is fully incapable of flowering and therefore does not reproduce. It is dwarfed. Only very young plants may make a full recovery from bacterial deprivation, although the growth height and leaf proliferation of older plants are improved after treatment of older plants with the specific bacterial suspensions. The bacteria display a life history. Although they do not form spores they are capable of dormancy for more than a year. They are motile, presumably swimming by means of typical rotary bacterial flagella, but they lose their motility and have a swollen ("bacteroid") shape when found in the mature flutes of the leaves called "leaf nodules."

The best known of the total of thirty species of *Ardisia* is *A. crispa*. All these and five species of the lesser-known genera bear regular leaf bacterial symbionts. DeJong suggested that this association originated from nearby soil bacteria, perhaps only once (monophyletically). The symbiosis is retained, he argued, because the plants have become dependent on growth-promoting substances released by the bacteria. The bacteria that form the fluted leaves of *Ardisia crispa* are so well integrated into the plant that they are simply taken as a characteristic that distinguishes the leaves of this species.

We suggest that this integrated bacterial symbiosis correlates both with leaf morphology and with the establishment of the new species and genus of *Ardisia*. Species differentiation between *Ardisia crispa* and the differing other members of the subgenus "Crispardisia" is directly correlated with the acquisition and integration of related but not identical bacterial symbionts.

Defensive Ciliates, Rosati's Revelation

Take another example. In the water along the shores of the Adriatic Sea thrive ciliates, single-celled protoctists, that clearly are members of the great family of ciliates called "heterotrichous." "Hetero" =

other, "trich" = hair—the name comes from the differing sizes and shapes of the motile protrusions on their bodies (Figure 3.2). Like other heterotrichous ciliates these have conspicuous "mouths" lined by rows of cilia called membranelles. The membranelles of the cell opening at the mouth sweep in bacterial food. The swimming ciliate is asymmetrical: the dorsal, or back, side is covered with fewer bundles of cilia (called cirri) than is the ventral, or front, side.

Six species are known in the genus *Euplotidium*. All are thought to have evolved from a common ancestor called *Euplotes*. *Euplotes*, a genus of heterotrichous ciliates that contains hundreds of species, is a cosmopolitan bacteriovore—it feeds on bacteria. *Euplotes* is far better known than *Euplotidium* but is similar to it in overall shape.

The distinctive feature of *Euplotidium* is a conspicuous band of knobby projections on its outer surface. As one zooms in to this band, studded with hundreds of bumps, it is seen to be composed of closely aligned bacteria. The bumps, which are clearly part of the ciliate and its description, are live bacteria. They have their own complex life histories. In stage one the externally connected or "ectosymbiotic" bacteria divide on the surface of the ciliate in typical "binary fission" fashion. That is, each bacterium (bump) divides into two equal bacteria offspring. In stage two each bacterium differentiates, changing to form a complex structure: On each bacterial bump an apical dome-shaped new structure appears that harbors DNA and protein. This complex DNA-protein dome has inside a conspicuous extrusive apparatus that is easily seen with a high-power microscope. A tightly coiled ribbon-shaped structure appears within the protein background. The ribbon surrounds a center object and a net of tubules. Even though they are inside a bacterial cell these tubules—long, hollow, and twenty-four nanometers in diameter—are surprisingly similar to the typical microtubules of eukaryotes. They are sensitive both to treatments known to impede the polymerization of microtubule protein into tubules (such as colchicine or cold temperatures) and to antitubulin antibodies.

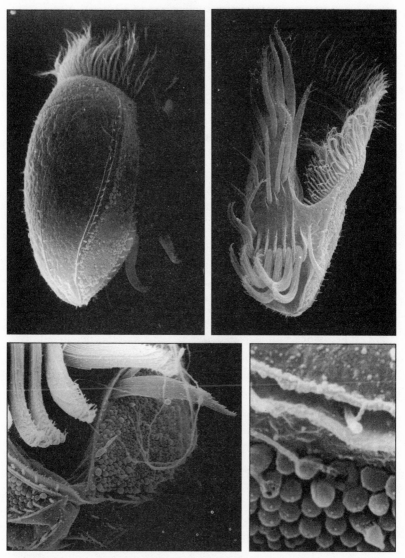

FIGURE 3.2   *Euplotidium* "Body Farms" Its Defense Organs

They are unquestionably microtubules, the structures that underlie the famous normal mode of eukaryotic cell division (mitosis), which are ubiquitous in nucleated cells and taken as a sign of their identity. The presence of such structures in bacterial cells is rare, almost unknown.

The transformation from stage one to stage two is fully correlated with the cell cycle of the ciliate in whose surface these bacteria are embedded. The ciliate detects signals in the Adriatic seawater that warn it of danger: potential predators in the form of a different ciliate, *Litonotus lamella*. The reaction is violent: the shooting out of the ribbon from inside *Euplotidium*'s surface bacterium. The forcibly extruded ribbon, or rather ribbons, since hundreds of bacteria shoot out dangerous ribbons at once, is forty microns long, nearly as long as the ciliate itself, whereas the coiled structure from which it was ejected is only about one micron. The two symbiotic organisms thus act as one: When the ciliate perceives itself to be threatened by predators or other disturbances, huge numbers of the surface bacteria shoot out their ribbons.

The association between the ciliate *Euplotidium*, in all six species of it, and the ribbon-firing bacteria is constant. All members of the genus bear these protective ectosymbionts. In the laboratory the ciliates can be deprived of their bacterial band. Without their bacterial protectors the ciliates swim, feed, grow, and reproduce by division normally. But the predator *Litonotus lamella* is always on the lookout for *Euplotidium*. Once deprived of its vestment of ectosymbiotic ribbon-shooting bacteria, *Euplotidium* becomes easy prey for *Litonotus*, which eats it immediately. We may say then that the bacteria on the surface of *Euplotidium* prevent its natural selection. We suggest that the acquisition of these bacteria and their integration as organelles by a *Euplotes*-like ancestor led to the new species *Euplotidium itoi*. Since all five other *Euplotidium* species all bear protective bacteria, it seems likely that the genus *Euplotidium* arose from *Euplotes* by microbial genome acquisition and integration. Today the ribbon-shooting bacteria of *E. itoi* are extracellular organelles: The ciliate cannot live in nature without them. The bacteria cannot be grown in culture unless they are attached to *Euplotidium itoi*. The bacteria depend on the ciliates to which they are attached for their very existence. Members of the genus *Euplotidium* exist only as symbiotic entities; the two organisms are not viable separately. We thus have a clear, presumably fairly recent example of the origins of species by inheritance of acquired genomes.

COMMUNITIES AND THE LOGIC OF ECOSYSTEMS

Members of the same species that live together at the same time in the same place belong to the same population. All organisms live in communities made of populations. All communities are made of more than two different species that live together in identifiable habitats. Such organization in nature precisely correlates with climatic, geographical, and other environmental factors. Taxonomy, the science of naming, identifying, and grouping of live beings, tends to ignore environmental correlates and is thus artificial, especially when applied to bacteria. Even so, in protoctists, fungi, plants, and animals, the most easily identified and named group level is that of genus and species. In spite of the zoological admonition that members of species be in the mating game and produce fertile offspring, species are distinguished, counted, documented, and named mainly based on their appearances, their morphology.

Although many different types of bacteria are recognizable, bacteria are far less stable in their characteristics than larger organisms. Bacterial "species" are so elusive and undefinable that the species concept does not apply to them. Some call the bacterial groups "strains," saving discussion of the details for later. Any bacterium can pass genes to any other. Restrictions on promiscuous gene flow, and thus the possibility of speciation, began in the lower Proterozoic Eon, about 2,500 million years ago, when the transformation from bacteria cells to consortia and communities led to integration and boundary-making and finally to the earliest eukaryotic cells. These cells, as we have explained elsewhere, are themselves symbiotic assemblages. Eukaryotic cell parts such as mitochondria and maybe even cilia and their microtubules originally evolved as free-living organisms. Symbiogenesis thus explains the origin of species in a double sense: first by bringing bacteria together in the ancient past to form the earliest organisms capable of speciation, and second, as we will see, by creation of particular new species by the incorporation of bodies and eventually of genomes.

CHAPTER 4

THE NATURAL SELECTOR

NIETZSCHE'S WILL TO POWER: REPRODUCTION

All organisms carry within themselves the chemical equipment for reproduction. Bacteria divide by twos: two becomes four becomes eight becomes sixteen. *Ad infinitum.* Foraminifera—large, shelled, ocean-dwelling protoctists—multiply by the hundreds. A single parent foram may generate 320 little swimming offspring in a few minutes. In animals and plants the mating game is often the only game in town. The ultimate drive, the goal of life, is to reproduce, with or without a mate. Friedrich Nietzsche described life's incessant tendency to have its own way, to create and expand, to view things according to a certain poetic interpretation that is then forgotten, becoming truth, as "the will to power."

Standing in opposition to this creative force is natural selection. Natural selection does not generate new forms, it does not innovate or produce. As its name implies, it only selects from among that which has already been created.

What then does the selecting in natural selection? Just as many modern evolutionists permit themselves an unscientific vagueness about the role of natural selection in evolution, they also remain vague about the identity of the natural selector. It is all too easy to wave one's arms and say "the environment selects, the fittest survive." What does "fit" really mean? What parts of the environment select? How far does the environment extend? Questions like these tend to be answered only in generalizations or in an ad hoc manner, case by case. A staunch resistance to any systematic effort to identify the agent or agents of natural selection takes place.

The simple but important assertion made in this chapter is that the natural selector is Gaia. Gaia, the biosphere, is best understood as the whole Earth's surface of interacting conditions and the biota, living matter, naturally organized into ecosystems. "Biota" or total biomass refers to flora, fauna, and microbiota taken together. Natural selection is one of the means by which Gaia, the self-regulating system, maintains itself as a dynamic but stable entity.

UNCHECKED POPULATIONS

Daily environmental constraints such as lack of water, crowding, and starvation prevent populations from the indefinite expansion of which they are capable. Since each population has specific energy, carbon, nitrogen, water, space, and other requirements that are never fully provided for by the environment, population expansion inevitably is stressed. Natural selection, a strictly subtractive process, eliminates all who fail to survive to reproduce for any reason. Those that remain, by definition, survive and tend to pass on heritable traits to their offspring. Since survivors retain traits most conducive to survival at given times and places, life on Earth retains a memory of its past. Living bodies store in their complex chemistry memories of past environmental limitations that they successfully overcame.

DARWIN'S ADMISSION

Why does a single termite harbor thirty different species of protists and over 200 distinguishable kinds of bacteria? Why, at the equator in the tropics where the sun rises each day at six o'clock and sets each dusk at six o'clock, where physical conditions are so uniform, do living beings so flagrantly differ? Why, on Galapagos Islands' rugged volcanic peaks, desolate in the ceaseless sunlight, so similar in size, in history, in physical predicament and geography, do the inhabitants vary so profoundly? "This has long appeared to me a great difficulty," Darwin, as quoted by Jonathan Wiener in *The Beak of the Finch,* wrote toward the end of *The Origin of Species*, "but it arises in chief part from the deeply-seated error of considering the physical conditions of a country as the most important for its inhabitants: whereas it cannot, I think, be disputed that the nature of the other inhabitants, with which each has to compete is at least as important and generally a far more important element of success."

Whether or not a newly hatched, or born, or germinated being grows, lives, divides, and survives depends in part, of course, on weather and geographical circumstance. But whether the new bit of life becomes food for the other, or whether it is a friend or ally in any great battle, or if it becomes the beneficiary of shade or warmth or suffers the wrath of any other depends mightily on the kindness or otherwise of strangers. No matter its nature, whether or not any being lives to reproduce depends on the other beings in its immediate surroundings, both its own and other species.

GAIA AS NATURAL SELECTOR

Gaia, the old Greek name for planet Earth newly reemployed as the name of a theory of the living planet, is depicted in Figure 8.1 (page 131) and defined here.

The term "biota" refers to all flora (plants), fauna (animals), and microbiota (fungi, protoctists, and bacteria), that is, is equivalent to the sum total of all of the biomass on the planet. Biota may be best understood as all living matter today. Biota begins and ends embedded in the biosphere—the approximately twenty-three kilometers from below the oceanic abyss to the top of the troposphere where life exists. Over thirty million types of organisms, species, and bacterial strains, descended from common ancestors. All interact. All produce and remove gases, ions, metals, and organic compounds. The metabolism, growth, and interactions of these myriad beings, especially in aqueous solution, lead to modulation of more than the temperature, alkalinity, and atmospheric composition at the Earth's surface.

Clearly then, Gaia refers to the sum of all the other living organisms, other than any arbitrarily chosen organism A embedded in its environment at the Earth's surface. The Gaian view of life includes the environment of all these other living forms as well. Gaia—in toto—is the natural selector of any organism, say organism A. Gaia, in general, is what prevents populations of organism A from reaching their biotic potential. Like all populations, from the anthrax bacteria in a warm lung to the New York rats in brownstone cellars, growth by reproduction that leads to more growth by reproduction will go on and on and on until it is stopped—by natural selection. The fact that Gaia naturally selects helps us understand how the Earth is an integrated living system.

CHAPTER 5

PRINCIPLES OF EVOLUTIONARY NOVELTY

Development of superpowers through chemical- or radiation-mediated mutation—theoretically giving rise to new species—is a phenomenon found mostly in comic books. Mutations are defined as differences between parent and offspring that may be transferred genetically to the next generation of offspring. They are either spontaneous or induced by treatment. H. J. Muller, by exposing fruit flies to X rays, helped discover the chemical basis of mutation. Many others have applied mutagenic agents, either radiation or chemicals, to many kinds of organisms with success. Heritable changes are documented but all who have tried to mutate to form new species have failed. Cancer drugs are mutagenic but they never lead to the making of new species. Various kinds of DNA base-pair changes and chromosomal mutations sometimes generate small heritable changes.

But mutation by itself usually generates sick or deficient life forms. Favorable mutational changes are always small. New mutations generate variations in members of the same species but the accumulation of mutations has never been shown—in laboratory organisms or in the field—to lead to crossing of the species barrier. Many attempts have been made to show the effects of mutations on evolution. Perhaps the most novel is that summarized by Barbara Wright in bacteria. Mutations given highly specific conditions are not random and they, in certain environments, are favored in bacterial descendants (Wright, 2000).

Unlike mutation, the rapid acquisition of new, highly refined traits by acquisition and integration of former strangers confers immediate selective advantages on protoctist, plant, or animal captors. Often the association begins as predatory: One organism attempts to ingest and digest the other, which resists. The subdued prey or undigested bacterium leads to a trapped population. The inheritance of trapped populations, especially in the form of microbial genomes, creates novel evolutionary lineages that display genuinely new strategies worthy of study. Of course random mutations occur—in both predator and prey—and of course they are important aspects of the evolutionary process. But mutations refine and hone. In eukaryotes, by themselves, mutations do not create new species or important positive inherited variation. Genomes, on the other hand, come neatly packaged with long histories of heritable virtuosities and synthetic tricks. They provide just what is needed for an organism to change drastically and yet remain coherent and viable. Good documentation in the science literature shows that mobility, food-making ability, or novel metabolic traits (vitamin production, extraction of nitrogen from air, detoxification of oxygen) may be appropriated in toto, like skillful workers of a previous business brought on board by an acquiring corporation.

And just as a marriage or corporate merger cannot be simply reversed, evolution by genomic acquisition is an irreversible process.

After integration of smaller genomes into those of the partner, the once-external genome can no longer be released back into nature. There comes a point in the relationship where the former free-living form cannot, by exercise of will, be rejected or "unacquired." Examples exist of the rejection of long-standing partners but they result in streamlined and altered new beings, not a reversion to past forms.

Lamarck was incorrect in saying that behavioral changes in the parent generation lead to inherited changes in the offspring. But dramatic new traits may be acquired in a single lifetime through adoption and subsequent integration of genomes. In our earlier work *Microcosmos* (1997) we described the only laboratory example of the origin of a new "species" we know about: the permanent and ultimately required bacterial infection of amebas studied by Kwang Jeon of the University of Tennessee. In certain cases behavior and body changes may be inherited when they are caused by the inheritance of a bacterial or viral genome. A clear example, the leaves of *Abutilon pictum* "Thompsonii," a plant variety whose spots are caused by a virus, is shown in Figure 5.1. In such cases Lamarck may not have been so wrong: The trait by itself is not inherited but the acquired genome that determines the trait is inherited. Genital rubbing and scratching are behaviors that promote transmission of intimate microbes, such as the syphilis spirochete or the protist *(Trichomonas vaginalis)* that causes vaginal itch. Sneezing spreads respiratory organisms and viruses. Light-seeking behaviors are seen in green worms that harbor algal symbionts but not in their white ancestors. The effect of symbiont acquisition on behavior is a theme worthy of investigation by serious evolutionists.

SOURCES OF HERITABLE NOVELTY

How are whole new genes acquired by any organism? Predatory bacteria penetrate their prey. Other actively feeding bacteria produce

FIGURE 5.1   Virus Variegation: *Abutilon pictum* (Malvaceae)

slime, whitish polysaccharides that become adhesive materials. Still other bacteria are gobbled up by protists. These behaviors lead to acquisition, attachment, and sometimes to fusion. The newest symbioses, like the bacteria *(Aeromonas hydrophila)* that feed between the cells or the ciliates that cruise the wet surfaces in search of bacteria that leak food in *Hydra viridis,* help us understand how deeper, more permanent associations began. A certain swimming protist in the intestines of termites is called *Streblomastix strix.* All *Streblomastix* are covered with sensitive, sensory hairs that recognize the acetate they ingest as food. The sensory hairs were revealed to be hanging-on bacteria. After they were removed by antibiotics, the *Streblomastix* were no longer attracted by the acetate food. In the evolutionary sense the protists were probably less likely to survive if their acquired, inherited bacteria sense organs were removed.

The membrane-bounded green plastids of algae were once tough and resistant food bacteria. Cannibalistic termites retain their nestmates' symbionts in their own intestines. The nematocysts—natural poison darts built from cell parts—in sea slugs are detriggered and borrowed from the coelenterates they eat, only to be redeployed against the slugs' own enemies. These examples not only show us how integrated symbionts tend to keep acquiring more symbionts; they show us how symbiotic history is evolutionary destiny.

Because it is nothing more than differential survival, natural selection perpetuates but it cannot create. What then generates evolutionary innovation? The list of sources of heredity change—mutation in the broad sense—is now growing rather long. Acquisition and integration of genomes is only a single entry. Novelty appears and accumulates by single base-pair changes in DNA, random or not. DNA mutations are easily inferred by direct study of amino acid sequences in proteins. This is because the amino acid sequence (in protein) is coded for by the base sequence (in genes). Perfect copying is not to be expected in a cosmos governed by thermodynamics's second law. Organisms also gain new hereditary traits by accumulation of viruses, or of plasmids or other short pieces of DNA. They acquire long pieces of new DNA, many genes at once, by bacterial mating and by legitimate sexual mating with distant relatives—that is, by hybridization. The non-genome-acquiring means of building up variety, that is, the acquisition of genes one at a time, are disproportionately important in bacteria. Taken together these means constitute a strong case that Darwin's dilemma is solved. Science now knows the major sources of evolutionary novelty. The task now is to bring this knowledge out of the dusty tomes of esoterica and make it better known.

How is new genetic material acquired? How much new genetic material can be acquired a gene at a time, or just a few genes—the ones desired? The new technology of DNA-sequencing lets us answer these questions directly.

The minimal heritable genetic change is a single base-pair change in DNA—from A-T to G-C (or from G-C to A-T). The maximal heritable change is the acquisition of an entire set of genes to run an organism (the genome) along with the rest of the organism

Table 5.1—Sources of Evolutionary Innovation

PROKARYOTES

Chromonemal
> single and few base-pair changes in DNA (point mutations)
> duplication of genes; acquisition of plasmids, viruses, prions, entire viral
>   genomes (Figure 5.1)
> recombination of genes and parts of genes
> loss of single or few base pairs, gene loss (deletion)

EUKARYOTES

Chromosomal
> polyploidization (including aneuploidy)
> translocation of chromosomes (meiotic and mitotic recombination)
> karyotypic fission (= kinetochore reproduction) in gamete precursor cells
>   (Chapter 12)

Genomic
> acquisition and integration of entire symbiont genomes
> loss of symbionts

in a healthy state so that the genome may have something to run. In between are many other ways in which organisms gain and retain heritable novelty. When the complete sequencing of the human genome was announced at the beginning of this millennium, many were quite surprised to learn that some 250 of the more than 30,000 human genes of our bodies have come directly from bacteria. These genes, long sequences of DNA that code for proteins, are as recognizably of bacterial origin as a feather is recognizably from a bird rather than, say, a shark's mouth.

How bacteria passed genes to people no one knows, but a good guess is via viruses. Bacteria are notorious for harboring viruses and moving them to new localities, such as to other bacteria.

The genome of the common yeast *Saccharomyces cerevisiae* has been fully sequenced, and it gave the scientists who did it a nice surprise. They chose *S. cerevisiae*—a single-celled fungus—as the representative of the fungus kingdom to sequence because this versatile little cell's life is tied to ours in many ways. This yeast makes dough rise, and therefore most baked goods—bread or cake or brioche—depend on it. It brews beer, therefore all beer with alcohol depends on happy growing conditions for *Saccharomyces cerevisiae*. It abounds in yogurt and other dairy products. It grows quickly and well under laboratory conditions and has been a favorite object of study in the investigation of fungal sex, fungal viruses, chromosome behavior, growth, and survival as well as spore formation. Each *Saccharomyces cerevisiae* yeast cell, as was well known, is "haploid," which means it has only one copy of each of its ten chromosomes. (We human animals are diploid, which means each of us has two copies of each of our twenty-three chromosomes; one set comes from our father, the other from our mother). What surprised everyone was that haploid yeast had two copies of nearly all the genes.

The yeast story adds another item to our collection of ways to gain new genetic materials: duplication. Every organism that has been studied has some detectable degree of gene duplication: a part of an older gene, an entire single gene, a set of a few genes, a chromosome's worth, or—as in yeast—nearly every gene in the cell's little body. Just as extra copies of manuscripts or instruction booklets free up the originals to differ from the copies, extra sets of genes have proved to be very useful as yeasts and other organisms have evolved to larger sizes and more complexity.

PART TWO

THE MICROBE
IN EVOLUTION

CHAPTER 6

SPECIES AND CELLS

*D*arwin claimed, and we were all taught, that evolution proceeds by gradual, nearly imperceptible steps. "The borders between species are as fluid and adaptable, as sensitive to changes in pressure, as the heaving of waves in a high sea," wrote Jonathan Wiener in *The Beak of the Finch*. "And winds can split the waves, as if splitting the mountain or sending a new mountain or new archipelago up above the rest. . . . A detail can make the difference. Even a detail that has no adaptive significance can make all the difference in the world. In other words, the origin of species can lie in the kinds of small subjective decisions and revisions that in our species come under the heading of romance."

Wiener, reflecting on the consensus among today's evolutionists, attempted to summarize a problem that Darwin himself faced with confusion and ambivalence: the relation between sexual and natural selection. Wiener's examples are two human beaks: first, the fatally attractive nose (attached of course to the face and body) of Cleopatra, who moved men like Julius Caesar and Marc Antony to actions that ultimately changed the population structure of the peoples of the Mediterranean Sea. The second was the nose of Charles Darwin himself, deemed by Robert FitzRoy, chief executive officer of the *Beagle,* to be that of a lazy man. If Captain FitzRoy had failed to hire the young naturalist as his companion on the *Beagle's* survey of the South American coast, the rest of us would not be pondering *The Origin of Species.* (Perhaps we would now be discussing some seminal work by Alfred Russel Wallace or T. H. Huxley instead.) Even the erotic whim of a British sailor enchanted by the pendulous breasts and straight sparse pubic hair of Solomon Islanders may have evolutionary repercussions, if his attraction leads to a new subspecies of Pacific islanders. So may a brutal hurricane, if only two black-skinned flat-nosed adolescents survive to repopulate the heights of the volcanic cones these marooned Polynesians call home. Sexual and natural selection are variations on the same theme: Only a tiny fraction of the varying many survive to propagate their kind.

The rub, the difficulty, the cause of the soul-searching and excess verbiage, is Darwin and his followers' assertion, or tacit assumption, or textbook "fact," that this kind of selection leads to new species. The heavy hand of selection can and does dramatically change the proportion of heavy egg-laying hens, grape-sized versus cherry-sized tomatoes, or long-beaked ground finches. But no one has ever shown that this process does more than change gene frequencies in populations. Intraspecific variation never seems to lead, by itself, to new species.

As a student in the 1960s Niles Eldredge, who later became the curator of invertebrate zoology at the American Museum of Natural History in New York City, sought direct evidence of speciation in the fossil record. He studied collections of wonderfully preserved Cambrian trilobites and looked for gradual transitions from one species to its descendant species. He combed the sediments in stratigraphic sequences in Morocco and upstate New York. He found some variation in size and shape of carapace from bed to bed, but never any clear trend indicating a slow transition from one species to another. Rather one species would continue with minor random variations for 800,000 years. Another would abruptly begin and overlie the first for 1.3 million years. The search for intermediate forms and gradual evolutionary change between the two species was always futile. The sedimentary rocks in which the glorious fossil record is embedded do not lie. They do not deceive. The record was punctuated, and the differences between species of extinct animals trapped in it were clean and distinct. The small variations within species indicative of changing gene frequencies would oscillate back and forth without direction (this was the "equilibrium" in "punctuated equilibrium"). The appearance of new species and genera and the loss of old ones by extinction was always discontinuous. (This is the "punctuation.")

Eldredge's trilobite conclusions, later reestablished by Richard Fortey and colleagues, were mirrored in Bahamian snails (by Eldredge's "theory of punctuated equilibrium" in scientific papers coauthored with Stephen Jay Gould). A discontinuous fossil record was seen in clams and dinosaur remains, and in the history reconstructed from bony plates of huge extinct fish that terrorized other marine beings in the Devonian seawater. Everyone who has contemplated the difficulties of reproductive success between a dachshund female and a German shepherd male dog has noted variation within a species. Yet no one who carefully studied his or her organisms in nature has detected gradual change of one species into another, either in live populations or in their fossil remains. Whenever investigators came close to the boundaries, the genuine gaps—between radishes and cabbages, donkeys and horses, or bonobos and chimps—remained. At least in easily recognized animals the species borders did not blur.

BACTERIA AS ONE SPECIES

Bacteria don't speciate; their different types are fluid and rapidly changing. Compared to plants, animals, and fungi, the metabolic repertoire of bacteria is vast. Metabolism in their protoctist descendants is more circumscribed but still more varied than the generalized metabolism of plants, animals, and fungi. Long before the appearance of fungal, animal, or plant life, choice, development, mating type, immunity, mineral formation, movement, photosynthesis, predation, programmed cell death, sexuality, and other "higher" attributes of life were already exquisitely refined in the bacterial world. We have only now begun to appreciate the dimensions of this kingdom's astounding metabolic repertoire and morphological diversity. (We have even made the case that in bacteria or their protoctist brethren there originated something like free choice—for which Christian de Duve, the cell biologist and Nobel laureate, accused us, in the pages of *Nature* magazine, of "biomysticism.")

A prodigious technical literature shows that bacteria are the main repository of evolutionary diversity. The entire animal kingdom employs essentially one mode of metabolism: the use of oxygen to respire organic food molecules. This is called heterotrophy. Plants employ two: heterotrophy via oxygen just as in animals, and carbon dioxide-fixing, oxygen-producing photosynthesis by use of the sun's radiation. This is called photoautotrophy. Bacteria, in addition to heterotrophy and photoautotrophy, have at least *twenty* metabolic modes that are not possessed (except via symbiotic acquisition) by either animal or plant. Some bacteria breathe sulfur or arsenic rather than oxygen. Others glow in the dark by production of "cold light"—the luminescing luciferin-luciferase reactions. Some bacteria thrive in foul-smelling communities where swimmers transform metallic ions in the waters to metal precipitates such as manganese and iron oxides or even gold. Certain bluish or green bacteria photosynthesize and expel oxygen in the light whereas in the dark they take up and breathe in oxygen as do animals and fungi. Others, of different kinds, use sunlight to photosynthesize in ways that preclude either the release or breathing in of oxygen—ever. Still different bacteria, called methanogens, change CO_2 and hydrogen into swamp gas. In the total absence of oxygen gas and food they make methane (CH_4), exactly the same gas we burn in "natural gas" stoves. The methane flows into crevices where still another obscure type of bacterium, a "methylotroph," burns it. Methylotrophs burn methane and derive energy from it in order to thrive and grow. Still other bacteria grow on deep-sea miniaturized light-based chemical reactions that clever scientists can neither explain nor imitate.

Understanding evolution's inner workings requires understanding the full range of life's possibilities. In the flexibility, cosmopolitan virtuosities, and sophisticated strategies of bacteria we can begin to appreciate evolutionary complexity in its great glory. Most bacteria disdain the company of humans. They reproduced, explored, conquered, gene-traded, indulged in their bizarre forms of sexual encounters, and eroded rocks long before any animal wandered the Earth. More bacteria by far orbited and landed on the Moon than did any men. That bacteria duplicate, transfer, digest, and in other ways lose and gain genes is an essential aspect of the evolutionary saga. The speed, volume, and antiquity of bacterial gene-trading activities underlie the evolution of all the rest of life on Earth. Humanity is a biospheric plume, an untested experiment that may disappear in self-consumption. The nucleated cell was a most spectacular outcome of bacterial machination. It, unlike humanity, is time-tested.

INCORPORATIONS:
WHERE SPECIATION BEGINS

By forming teams that integrated to form the first nucleated cells, bacteria began the speciation process. These new cells, some of them, transformed into protist descendants. They evolved new sets of capabilities that led them to leave the bacterial realm. Speciation as a process began here, in the lower Proterozoic world, perhaps 2,000 million years ago.

Protoctists extended the bacterial world. Like the embryo-forming animals that succeeded them, they evolved feeding and social strategies. Here in the protoctist world, the first engineers and designers created structures like delicate silica boxes or calcium carbonate hunting platforms. Others invented agriculture. Single-celled foraminifera, representing some 60,000 distinctive species and comprising relatively huge cells, farmed and trapped algae they expelled along body tracks and stored in stalls in their shells. The well-trained algae take in the sunlight and manufacture food for everyone nearby during daylight hours. Then, snug inside the foram shells at night, the algae breathe in oxygen that they produced during the day. Some fussy relatives of these foram farmers make their new shells from a multicolor mix. These "agglutinators" plaster tiny stones to their bodies. Other forams even pack minerals together to construct lookout towers. They climb atop the towers and hunt, preying on animals such as rotifers and crustaceans far larger than themselves. Protoctists include large "water neithers," organisms neither animal nor plant, such as kelps, 100-meter-long photosynthesizers still with us today. The fossil record for foraminifera and diatoms is vast even compared to shelled marine animals such as clams and trilobites. But descriptions of protoctists are sparse because of the paucity of paleontologists and protistologists alive to describe them. We could learn from their miniature technologies and incredible range of symbiotic associations—if we wished.

CELL EQUATIONS AND SPECIES DEFINITIONS

Some 450 million years ago, bacteria, protoctists, and animals were joined by fungi and plants. Whereas plants belong to our world of the easily visible, most fungi, such as yeasts and molds, fit the description of microbes. The mushrooms and certain large and complex molds are easily recognizable to the unaided eye but even they, like all microbes, are best seen and understood with microscopes. Most fungi are tiny white scum in the unmagnified image. Professional and occasional amateur mycologists, whose numbers dwindle annually, are in practice the only people who distinguish one species from another. Some fungi wrap their stringy bodies and become spring-loaded traps for nematode worms, which they squeeze to death as a python squeezes rodents. Others enjoy bizarre sex lives. Not male and female but thousands of genders, for example, are known in *Schizophyllum commune* and other woodland mushroom-like species.

Microbes, (prokaryotic bacteria, eukaryotic protoctists, and fungi), in short, evolved to act in ways that extend far beyond the relatively uniform practices and lifestyles of animals and plants. To understand the machinations of the evolutionary process in animals and plants, we must also acquire an awareness of the microcosm and its denizens' talents, urges, exudates, and gender and reproductive peculiarities. Microbes have uniquely capable complete genomes. They, not selfish genes or combative male mammals, are the engines of evolutionary change.

The integrated genomic systems of the major kinds of life in short-hand equation form are shown in Table 6.1.

Table 6.1—Cell Genome Equations

Origins of groups from 2,000 to 541 million years ago
(PROTEROZOIC EON)

"Hypersex"[3] or
the making of the mitotic
—but meiotically sexless[1,2]*—single cell*

				minimal number of component genomes	
Components					
Thermoplasma (Archaebacterium)	+	***Spirochaeta*** (eubacterium)	=	**archaeprotist (2)** (with karyomastigont)	⎫
Archaeprotists (amitochondriate eukaryote)	or or +	***Paracoccus*** (α proteobacterium) **Bdellovibrio** (δ proteobacterium)	=	**"protozoa" (3)** (aerobic eukaryote)	⎬ Protoctista
"protozoan" (aerobic eukaryote)	+	***Synechococcus*** (cyanobacterium)	=	**"alga" (4)** (photosynthetic eukaryote)	⎭
"protozoan"	+	**"BLO"** (***Burkholderia***)	=	**fungus (4)** (asco- zygo- or basidiomycote)	
"alga"	+	**mycosome**[4] (yeast)	=	**plant (8)** (bryophyte or tracheophyte)	⎫ embryo
"protozoan"	+	(carbonate-precipitating eubacterium or other)	=	**animal (4)** (parazoan, mesozoan or metazoan)	⎬ formers

[1] Sex: Formation of new individual with genes from more than one parental source.

[2] Meiotic sex: routine fusion of bodies, cells, and nuclei (doubling) followed by routine reduction (halving) by meiosis. Characteristic of fungi, plants, animals, and many protoctists.

[3] Hypersex: permanent fusion of unlike ("differently named") organisms to form a new higher taxon, e.g., mitochondria acquisition by anaerobic eukaryotes to form aerobic protists or the sulfur-oxidizing bacteria acquired in the origin of the deep sea tubeworms (*Riftia pachytila*).

[4] Peter Atsatt, University of California, Irvine; see Atsatt, 2003.

CHAPTER 7

HISTORY OF
THE HERITABLE

Genomes, the sum of all genes in an individual, do not exist in isolation in nature. They are always embedded deep inside cells. The minimal genome of a free-living bacterium already sequenced has about 500 genes whereas, the genomes of plants and animals can be one hundred times as large—50,000 or so genes. For two different types of genomes to merge and form a new one, the organisms themselves must have a reason to come together. Reasons vary. Organism A may find B delicious, and try to swallow B. Alternatively, organism A may require the chemical form of nitrogen excreted in the waste of B. Organism A may simply bask, at first, in the shade provided by B—or B may sequester the alkaline moisture that exudes at dawn from the pores of A. These are ecological issues with many subtleties, but they underlie the transfer and eventual merger of microbial genomes to larger forms of life. How has the creativity of the microbe been transferred to plants, animals, and other large forms of life? How often has microbial creativity been acquired by larger life forms? Many times and in many ways. The maligned Lamarckian slogan "inheritance of acquired characteristics" need not be abandoned, just carefully refined.

Under stress, different kinds of individuals, of very different origins and abilities, physically associate. With continued and predictable stress, cyclical and seasonal, these acquaintances become intimate and extend beyond a single encounter. To become significant to the evolutionary process, the former strangers must interact frequently enough to form a stable, unique relationship—and ultimately a permanent or at least deeply seasonal affair. Put succinctly, in cases of importance in evolution, associations lead to partnerships that lead to symbioses that lead to new kinds of individuals formed by symbiogenesis. At any time the association may dissolve, the partners may change or even destroy each other, or the symbionts may be lost. Outcomes that involve very different live organisms are not fully predictable, and terms like "cost" and "benefit" are not very useful. Associations may tighten so much that literal incorporation occurs. From casual association and unavoidable metabolic exchange (where the waste products of one being, in an open thermodynamic system, become food or protection or lubrication to another) grow new mutually incorporated bodies. Former independents can be recognized as components of corporate mergers who have not lost all vestiges of their earlier independent state. The green chloroplasts of leaves still divide. The beating fringe of spirochetes on *Mixotricha* still swim—but they don't swim away. After long periods of complementary living, intimate association or metabolic dependency, the strange bedfellows often fuse their genetic systems. This is the last step in genome acquisition, the key to the "inheritance of acquired genomes." The once separable and differently named partners now become a new entity, a new individual at a larger level of size and complexity. Many such fusions have been documented in all five kingdoms of life.

The most interesting to us are those partnerships that spurred discontinuous and conspicuous evolutionary innovations so immense that they left clear evidence of their histories in the fossil record. Such outstanding examples include coral reef animals and their algae; photosynthetic clams; sulfide-oxidizing, two-meter-long tube worms in the abyss; urbanized termites; and grass-munching cows.

Variegated plants attract us. Our eye is drawn to colors spotted, blotched, striped, or patterned on leaves, fruits, flower, or stems. The reasons that a plant displays a variegated pattern are many but virus infection is certainly one. Viruses, in principle, are too small to function on their own yet their genomes enter plant, animal, protoctist, bacterial, or fungal cells and co-opt the metabolism of these hosts. Unsightly spots can become lovely decoration when specific viruses are introduced to new plants by grafting of the virally infected branch or stem to an uninfected plant. The viruses, for example, in hybrid tulips or certain *Abutilon* species (*Abutilon pictum*) cause stripes or beautiful bright yellow and cream patches. The viruses do not sicken the plants at all and people select and propagate these plants for their aesthetic virtues. Here we have another case of genome acquisition as source of heritable variation—caused by a virus within a species and maintained in the population by gardeners (Marcotrigiano, 1999, Fig. 5).

MEET THE BEETLES: HEDDI'S WEEVILS

The weevil, the scourge of neolithic man as his grain gardens began to expand, is with us today. This tiny beetle, and there are many kinds, lives entirely in Gramineae or Poaceae (members of the grass family). The latter are the seeds of the global expansion of our species: wheat, rice, corn, barley, and oats. The best-studied of these insects is *Sitophilus oryzae*, called the rice weevil but perfectly able to destroy stored grain of other kinds as well. It completes its entire life

history inside a grain of rice or berry of wheat. Some feel these beetles evolve with us. Apparently this is because they discard the genes
they no longer need as we provide them with the gene products
they no longer have to make for themselves.

For seventy years the presence in these weevils of bacterial tissue
(once called the mycetome as it was thought to be fungal) has been
under investigation. A specialized larval organ called the bacteriome
at the apex of the female's ovary is packed with bacteriocytes. These
swollen beetle cells harbor an enterobacterium that shares 95 percent genetic similarity with the common colon bacterium, *E. coli*.
Found only in three locations in the insect—as the larval bacteriome in both males and females, as the female ovary bacteriome, and
in the eggs (the female germ cells)—this bacterium, like our mitochondria, is maternally inherited. This implies that both male and
female infant weevils inherit the cells that will make up the bacteriome from their mothers. The sperm make no contribution. What
do the bacteria provide? At least part of the answer, wrested from
nature through painstaking work by many investigators, is: B vitamins. The bacterial genome is replete with genes that code for these
crucial nutritional supplements, which include riboflavin, pan-
tothenic acid, and biotin. When the bacteria are eliminated from
the tissues by antibiotic or other treatment, the weevils' growth rate
is markedly diminished. Bacteria-deprived weevils mature later than
symbiont-packed ones. Most importantly, because their supply of
mitochondrial energy is impaired, the weevils that lack their normal
bacteria cannot fly—an apt metaphor for the importance, in nature,
of symbiogenesis.

Weevil experts like Abdelazziz Heddi and his mentor, Professor
Paul Nardon, at the Institute National des Sciences Appliquées in
Lyon, France, call the highly integrated bacterium SOPE (*Sitophilus
oryzae* Primary Endosymbiont). From China to Guadalupe to their
own Lyonnaise backyard, to Italy where Umberto Pierantoni
(1876–1959) first discovered it, all strains of *S. oryzae* weevils bear the

same huge population of this bacterium in the same pattern. Perhaps the acquisition and integration of the symbiotic bacterium into the weevil's metabolism coincided with the origin of the species *oryzae.*

The plot thickens: Heddi and his colleagues in the late 1990s found a different kind of bacterial associate in weevils. Of some two dozen strains worldwide, representing three species of *Sitophilus* (*oryzae, granarius,* and *zeamais*), 57 percent harbored an additional bacterium in their tissues. Unlike SOPE, when the weevils were "cured" (by antibiotics) of the second bacterium, no effect was seen on the beetles' physiology, nitrogen nutrition, or flight. Rather the new bacterium, which resembles the *Wolbachia* strain, was far less regular in its habits. The new bacterial symbiont is found in variable numbers all over the insect's body with one notable exception: It is rampant in the germinal (sexual) tissues, especially where it can interact with developing sperm nuclei. Some investigators suspect that the new bacterium adheres to the peculiar chromatin-binding proteins of the sperm-producing tissue. Whatever the details of its action, this bacterium impacts the fertility of these prolific grain-eaters. When males that carried the *Wolbachia*-like bacteria in their spermy tissues were mated with females that lacked it in any tissues, fertility was dramatically decreased. The reciprocal cross, females replete with these bacteria in many tissues with males that lacked it, also depressed fertility but less so. The number of progeny was maximal—to the great advantage of the opportunistic bacterium—when the *Wolbachia*-like bacterial symbionts were present in both genders of the mating pair. The effect of the symbiosis is thus to promote reproductive isolation. Matings of uninfected weevils are of course fertile, matings where both partners are infected are fertile, but mixed matings are less so. Reproductive isolation is of course one of the key elements of speciation.

The concept that reproductive isolation, and therefore incipient speciation, might be induced by the presence of microbial symbionts is not new to the biological literature. The idea was well articulated by

Theodore Dobzhansky and his colleagues, who studied *Drosophila* in population cages. They mated fruit flies that had been separated and subjected to differing temperatures, cold and hot, for some two years. The matings had been fully fertile but now crosses between hot-raised flies and cold-raised ones were less than fully fertile. The probable cause of the decrease in fertility was the presence of mycoplasmas, wall-less bacteria, in the cold conditions and the loss of that tissue invader at the higher temperatures of the warm-incubated population cage. Nardon, Heddi, and many other authors have documented this observation: If male and female of the same species of insect both carry the same bacterium in their tissues, their coupling produced normal fertile offspring. Trouble begins, as with the rice weevil, when one gender carries microbes that the other lacks.

Dobzhansky and the others, though correct, were never explicit. They noted the presence and absence of bacteria and the effects on fertility, but they never raised this observation to the level of a general mechanism promoting speciation. When mycoplasmas or proteobacteria were acquired by one gender of insects and prevented fertile sexual outcomes unless the mate also carried the new microbe, "reproductive isolation" ensued and speciation followed. The attraction of bacteria to reproductive tissue in both females and males, especially in insects, was well documented by Paul Buchner (1886–1978) and his successors.

We again define species as follows:

Two live beings belong to the same species when the content and the number of integrated, formerly independent genomes that constitute them are the same. E. Mayr's species concept of 1948, which states that organisms may be assigned to the same species if, in nature, they mate and produce fertile offspring, becomes one example of our general rule. Mayr's concept is especially applicable to animals, who daily eat bacteria that may pass from the intestine to take up lodging in the gonads and other fatty reproductive tissue. The tiny strangers may swim through their tissue, find a niche, stay, and influence the future of what may become a new species with a whole new set of useful genes (the foreign genome from the talented but now lazy bacterium).

MICROCOSMIC DIALOGUE: PARTNER INTEGRATION

Darwin's question about how species originate may be rephrased as: "What is passed from parent to descendant that we detect as evolutionary novelty?" A straightforward answer is, "Populations and communities of microbes." Populations, remember, are individuals of the same type living at the same time and place, yet communities prevail in nature. When members of communities, different types of life thriving (or merely surviving) at the same time and place, fuse and transfer genes among themselves, new, more complex "individuals" evolve. The new, large, more complex individuals are invariably assigned new species names by biologists.

The stories of how microbes tend to physically join each other, and of their multiple interactions among themselves and with larger associates, have been told many times in the specialized language of the sciences. Inevitably these stories are poorly known, in part because the sciences themselves are so fragmented. Even those of us who understand how much already is known about the origins of species are limited to work on our own tiny discoveries, usually one species at a time. Academic biology departments once joined the zoologists with the botanists. Plants and animals were studied together. Falling through the cracks at liberal arts institutions, the study of microbes was largely shunted to its own departments, called microbiology, all of which resided in medical or agricultural schools. The main concern with understanding microbes—as disease agents or food contaminants—was in order to kill them.

Since the 1980s many biological science departments have splintered further into molecular versus organismal biology, a move that exacerbates misunderstanding. The relevant information on species origins is scattered across more than a dozen fields, each with its own esoteric language or languages. Identical organisms are sometimes partitioned into distinct disciplines. Cyanobacteria, for instance, are studied under phycology (or algology, a branch of botany) rather than bacteriology because they were once misnamed "blue-green algae." Biochemistry, cell biology, geology, invertebrate zoology, metabolism, molecular evolution, microbial ecology, nutrition, paleontology, protistology, sedimentary geology, and virology are all relevant to deciphering the origins of species. Most of these fields, black boxes to the public as well as to graduate students, remain mysterious even to many scientists who practice evolutionary biology today.

Many of Darwin's fashionable but misguided followers habitually misinterpret even the parts of science they know well. The revelation of much science beyond his century, extended by molecular biology and paleontology, is entirely consistent with Darwin's great insight. But that revelation shows that the luxuriant living diversity surrounding us *did not* evolve gradually, as the students of the fossil record so vociferously tell us. Precious little evidence in the sedimentary rocks exists for small steps that connect one species gradually to its descendants. The "traces of bygone biospheres," in Vladimir Vernadsky's immortal phrase, proclaim the opposite. Punctuated equilibrium is there for all who take the time to see it. The discontinuous record of past life shows clearly that the transition from one species to another occurs in discrete jumps. In trilobites, snails, seed ferns, horses, lungfish, sharks, and clams, evidence abounds for punctuated change. In spite of Darwin's own protest of an incomplete record and his claim of a "passage" of one form to another, for example in pigeons, barnacles, and dogs, animal and plant life has evolved in microbe genome-sized steps. What appears as magic, "irreducible complexity" or "grand design"—the retinal image of the vertebrate eye, the bumblebee's flight, the song of the ocean-traversing humpback whale or of the Three Tenors (Luciano Pavarotti, José Carreras, Placido Domingo)—is the legacy of repeated interactions. Metabolic dialogue and physical proximity led to incorporation, accommodation, and rearrangement. Familiar organisms are not simply or merely individuals; they protrude from the microbial underworld.

How then do independent, separate organisms fuse to form new individuals? How do those new individuals retain and perpetuate their composite selves? What do we mean by "acquisition and integration" of genomes? The answers to these questions are known in a few specific cases because the independent organisms of different types can be located in nature; intermediate forms (less and then more integrated) can be recognized; and the history of the composite can thus be reconstructed.

Ivan Emmanuel Wallin (1890–1967), a professor of anatomy at the University of Colorado Medical School, wrote *Symbionticism and the Origin of Species* (1927) when he was still quite young and lived in New York City. Wallin was a lone voice crying out that symbiosis is crucial to the evolutionary process. He invented his own name for symbiogenesis with microbes, calling it the formation of "microsymbiotic complexes." He dubbed the process of symbiotic integration in evolution "symbionticism," not realizing that rich literature on the same concept had been developing in Europe since the 1860s. He never knew, for instance, the work of Konstantin Merezhkovsky or Andrey Famintsyn. The European literature, published in Russian, French, and German, eluded Wallin because, like most scientists from the United States, he read only English. Since the late nineteenth century in the European scientific world, acquisition and inheritance of microbial symbionts had been called "symbiogenesis." Darwinism in general was disdained. Natural selection, it was claimed, from the beginning could not produce "good" evolutionary changes. Symbiogenesis was crucial to the generation of biological novelty. The Russian literature, interpreted by Liya N. Khakhina, a historian of science, was finally made available in English in 1992. It took two generations of scholars to summarize the great literature of the Russian botanists. Today it seems that the literature is ignored for just this reason. Old literature by Russian botanists lacks any market appeal in the Anglophone world. Nevertheless the Russians (primarily Famintsyn, Merezhkovsky, and Kozo-Polyansky, each with different emphasis) present the same thesis as Wallin: Symbioses and their consequences, especially microbial symbioses, generate evolutionary change.

Wallin first developed a theory of "prototaxis" to answer his question of how these tightly integrated symbioses could ever begin in the first place. He realized, as those of us who work in this field of science do, that all organisms are products of their highly specific history. Prototaxis, Wallin proffered, preceded symbiosis in any association of more than a single type of life. By "prototaxis" Wallin meant the innate tendency of one kind of cell or organism to respond in a specific manner to another sort of organism (Wallin, 1927, p. 8).

Examples include the tendency of the mouse to flee from the cat, the shark to swallow the fish, the rabbit to dine on lettuce, or the fly to lay its eggs on the bloody exposed muscle tissue of a recently dead boar. Prototaxis, as these examples show, may be negative or positive—associating or disassociating. We know them from experience. Prototactically we flee from the wasp nest and swim toward the water lily patch. All cases of symbiogenesis begin with prototaxis.

Photosynthesis, the natural chemical genius that converts light and carbon dioxide in air to energy and food, is entirely bacterial. All photosynthetic organisms, either bacterial themselves or descendants of bacteria, leak organic compounds from the prodigious bounty of their harvested light. The prototactical behavior of many nonphotosynthetic forms of life toward the photosynthesizers is simple to comprehend. Motility, whether by swimming, crawling, gliding, or creeping, serves to insure that the hungry being incapable of photosynthesis remains in the well-lit zones mandatory to the photosynthesizer. The prototactical tendency of "heterotrophs" to either absorb the products of photosynthesis or to ingest the photosynthetic organisms themselves has led to thriving communities of feeders in the sunlit surface zones of marine and fresh water. The secondarily photosynthetic organisms—those that acquired photosynthesis by ingesting the correct bacterium but failing to digest it—include all the photosynthetic eukaryotes. No alga or plant ever evolved photosynthesis on its own. All shared some ancestor—recent or remote—that ate but failed to digest a green or red or greenish blue bacterial photosynthesizer. Prototaxis, in this case, is the tendency toward hunger on the part of the eater and toward resistance to digestion on the part of the eaten. Starvation in the light and resistance to digestion, in short, have led over and again to permanently pigmented photosynthetic organisms: Algae, lichens, plants, green worms, green hydra, brown corals, and giant clams *(Tridacna)*—whose immense shells remain open to insure the focus of the light on the chloroplasts of their symbiotic algae—are only a few of the many examples upon which our lives depend. Prototaxis, these organic tendencies, may be early versions of the sort of purposefulness that, when we find it in ourselves, we call "conscious choice." But all beings have a relationship with time that is more complex than simple duration: They are oriented by their actions, associative and disassociative, to future consequences.

Symbiotic partners are integrated on many levels. The first and the most superficial is behavior. Before any permanent association begins the potential partners must be in the same place at the same time. What brings them to the sunlit water hole, the tree tops, the darkened cave, or the fog desert? Ancestry and contingency determine an organism's immediate behavior. Some stable symbiotic partnerships are integrated only on the behavioral level. Many, however, proceed to more intimate levels of integration: from behavioral to metabolic. Frequently the metabolic product, the exudate or the waste, of one partner becomes the food of the other. Probably all the green animals that have been studied (such as *Convoluta roscoffensis*, the flatworm, or *Hydra viridis* in ponds) and all the lichens are integrated at this level. The "photosynthate," that is, the sugars or other products of photosynthesis made by the green partner, leaks out to become food for the engulfer. The photosynthate may be released directly into the recipient's body. The eagerly feeding recipient inevitably urinates, defecates, or in other ways produces its own, often nitrogen-rich waste that literally feeds back to become a necessary nutrient for the photosynthesizer. Metabolic-level integration like this is quite common. When partners have a long history of association their integration is no longer easily reversible; their individuality melds. The malarial parasite *(Plasmodium vivax) must* derive its nutrition from the bloodstream of its victim. Here protist and mammal *(Homo sapiens)* are clearly integrated at the metabolic level. Lichens are also clearly metabolically integrated. The products of photosynthesis flow from bluegreen or green to the translucent partner. What is seldom realized is that in tight associations the metabolites flow in both directions. The animal or fungal partner also releases materials to the photosynthesizer. Symbioses are two-way exchanges. The kindness of strangers, the metabolic flow of gifts, makes them less strange and, ultimately, part of a single, co-dependent biological self. See Brodo, et al., 2001.

Metabolic associations can integrate even more tightly: The next level of partnership is called "gene-product association." Proteins (or even RNA molecules) of one of the partners are required for the other's functioning. A fine example of gene-product integration comes from bean and pea plants. If you dig up a clover, a vetch, or a string bean plant you may see on the roots little pinkish swellings. These are the nitrogen-fixing nodules. Inside the bumps a certain type of bacteria reside. Once a swimming variety of normal-looking rod-shaped bacteria, they have all converted to swollen "bacteroids." These over-sized hole-ridden "bacteroids" are unable to divide and grow ever again. The root nodule's pinkish color results from a protein called hemoglobin—the same sort of protein, although different in detail, that makes our blood red. In our blood the hemoglobin carries the oxygen from our lungs to our other tissues that need it. In the bean and pea plants, however, hemoglobin moves oxygen from the site of the nitrogen-fixing enzymes because oxygen gas will poison these enzymes. The crucial point, from the vantage point of symbiosis between pea plant and nitrogen-fixing bacterium, is that the hemoglobin molecule itself is chimeric. Whether in a bean or a human, hemoglobins have two components. One is the protein (the globin). The other is the part that gives the color. This, a much smaller nonproteinaceous molecule, is called the heme. In the plants with nodules, the heme is manufactured by the bacterial partner while the globin protein is a product of the plant. This is a story of gene-product integration in a symbiosis. The final product, the fully formed hemoglobin molecule, does not exist without the contributions of both bacterium and plant.

The greatest intimacy among partners involves integration at the genetic level. When a gene of one organism enters and remains with the genes of another, for example when a gene is passed from a free-living bacterium to the nucleus of a plant, integration is complete. No greater intimacy is known than the permanent harboring of your partner's genes. By the time this level of intimacy occurs, the chance is great that behavioral, metabolic, and gene-product integrations are already in place. When partnerships persist, they are intimate and share long histories, and many factors preclude their dissolution. Often they can never return to the status quo ante.

Turn over certain starfish off the coast of California and you will see a whole set of black little worms crawling at the starfish's centrally located mouth. Much food is gobbled up by the mouth but the worms are not harmed or ingested. Place the worms at some distance from the mouth and they will wriggle right back to the starfish's opening. Starfish are aggressive. The worms normally feed on leftovers and fragments right at the fearsome danger zone. This symbiotic association between worm and starfish is only behavioral. Two substances produced by the starfish affect the worms. One (destroyed easily by heat and therefore probably a protein) makes the worm wriggle quickly. The second, a heat-stable compound, attracts the worm so that it swims in the direction of the open mouth. Together the two compounds keep the worms at their feeding grounds. There must exist even greater chemical complexity, because the starfish doesn't eat the vulnerable worms at all but eats its similar relative and just about everything else edible in its mouth's vicinity. Probably some inhibitor compound discourages ingestion of these particular black worms, which, in their frenzied feeding on the offals, keep the mouth area clean and tidy—they provide a kind of natural oral hygiene.

Loose, behavioral associations, like this one between starfish and worm, may be stable indefinitely. We find these stable, loose associations in certain situations, such as cold winters that produce alternatively wet or frozen and dry ponds, in which the unassociated organisms leave more offspring separately than they do together. At other times of the year, the symbiotic pairing may be favored. The result is that the behavioral symbiosis persists but does not grow more intimate. The complete, irreversible integration of two different beings to form a new one will occur if at all times the physically associated organisms leave more descendants than do the independent unassociated ones. The hundreds of mitochondria, for example, in each of our cells, never leave these cells. Why? Because the world of animal tissue is full of oxygen and requires a flow of oxygen to the cell at all times. In the oxygen-rich Proterozoic, the cells that retained their mitochondria throughout their lives must have been "naturally selected" over those that occasionally let their mitochondria return to the bacterial world. Hence one must always ask how the partners are integrated, if they are always integrated, and what environmental conditions influence their integration. To substitute these sorts of details of metabolite flow and gene-product transfer between intimate former strangers

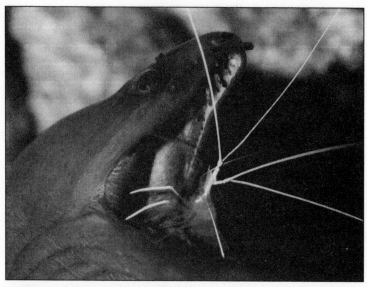

FIGURE 7.1   Scarlet Cleaning Shrimp *Lysmata grabhami* in Green Moray Eel's Mouth *(Gymnothorax funebris)*

with neodarwinian terms like "cooperation," "cost," or "benefit" is absurd and exemplary of the fallacy of misplaced concreteness. Such terminology precludes real understanding of the inevitably rich and complex evolutionary past of the symbiotic world that made animals, plants, and their nucleated planetmates. Like a sugary snack that temporarily satisfies our appetite but deprives us of more nutritious foods, neodarwinism sates intellectual curiosity with abstractions bereft of actual details—whether metabolic, biochemical, ecological, or of natural history.

FUNGAL FARMERS

Since the Mesozoic Era, probably as long as 100 million years ago, termites have lived in felled trees where they have eaten and nested in the wood. All termites are insects, in the animal phylum of arthropods, and are members of the order isoptera (*iso* = equal, *ptera* = wings). Most entomologists agree that the group, now 6,000 species strong, evolved from wood-eating cockroaches like the still-living *Cryptocercus punctulatus*. This idea is logical for two reasons. Roaches precede termites in the fossil record, and the few wood-eating roaches alive today still harbor hindgut communities of protists and bacteria obviously related to those found in the intestines of the "lower" (that is, wood-eating) termites (Figure 7.2).

Indeed the presumed most ancient ancestral-type termites still alive today are placed in a roach-like termite family called Mastotermitidae. Species of *Mastotermes* lived abundantly worldwide 5 million years ago, in the Eocene and Miocene, but today only one remains, *Mastotermes darwiniensis*. These relics of a bygone age still form huge colonies in the hot northern Australian port of Darwin. Rhinotermitidae, the family of subterraneans, the damp-wood-eating Hodotermitidae, and those who both eat and nest in dry wood, the Kalotermitidae, complete the list of "lower" isopteran families. All species of "lower" termites that are capable of the digestion of wood within their swollen paunches always, in all cases, are entirely packed with microbial symbionts, both eukaryotic and prokaryotic. (See Figure 7.3.)

All the other termites, by far most of the taxa, are considered "higher." Higher, or "later evolved," only means the ancestors of these termites abandoned the fascinating symbiotic hindgut wood-ingesting microbial nymphs and have evolved other more eclectic food-gathering strategies. Some so-called higher termites broadened their taste; they enjoy rich organic diets. Others remain totally restricted to the ancient way of wood-eating; lignin and cellulose are their only carbon sources. They survived their own paleolithic past but still they feed on wood of all kinds. They invented an agriculture where they cultivate a wood-digesting partner—a fungus (or perhaps many similar fungi) with the logical name *Termitomyces*.

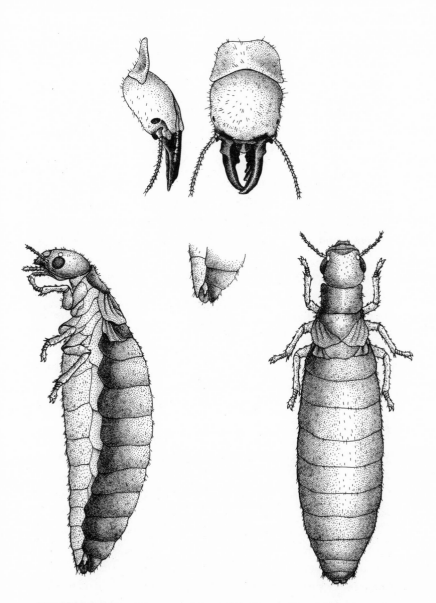

FIGURE 7.2 *Pterotermes occidentis* (Similar to *Heterotermes tenuis*)

"Higher" social termites domesticate these fungi, which they sow
and reap.

Mycologists blush and perhaps genuinely should be embar-
rassed at their clumsiness in growing pure cultures of *Termitomyces*
compared with the brown thumbs of their far more experienced termite farmers. Busy higher termites loosen and prepare the soil, fertilize it, prune unruly fungal threads called hyphae, and weed a
multitude of unwanted fungal growth and other debris. Finally, at
the end of the season, they harvest their crop. Even the most gifted
mycologist can not imitate the termite's skill in the growth of *Termitomyces* under cultivation!

"The effect of pruning and cropping is best seen by comparing a
well-tended fungus garden and the same fungus grown in laboratory
culture," wrote our friend Elio Schaechter (his full first name is Mose-
lio) in his charming book, *In the Company of Mushrooms* (1997). "In
artificial culture the fungi produce long filaments [hyphae], whereas
under the care of the termites the tips of the fungi become rounded
into clubs and spheres [staphyla], shapes not seen elsewhere. Abandoning a garden results in the outgrowth of the fungi, they 'go to
seed.'" (He means they begin to make reproductive structures called
spores.)

Apparently mycophilic peoples and tropical termites completely
agree about which species of fungi are delicious and worth the effort
to tend.

Millennia, indeed millions of years, of practice and evolutionary
selection have led to huge farming colonies of termites that live in
mounds called termitaria. The cities of *Macrotermes natalensis*, a
southern African mound builder, typically support one or two million inhabitants in each large air-conditioned conical structure.
These tower three meters above the Earth's surface. The nest itself,
underlain by a spacious cellar with vertical struts, is supported on
pillars. Approximately two meters wide, it is positioned so that two-
thirds of it lies below the surface of the ground. The "nest," the place
in the termitarium where the termites and their thriving gardens

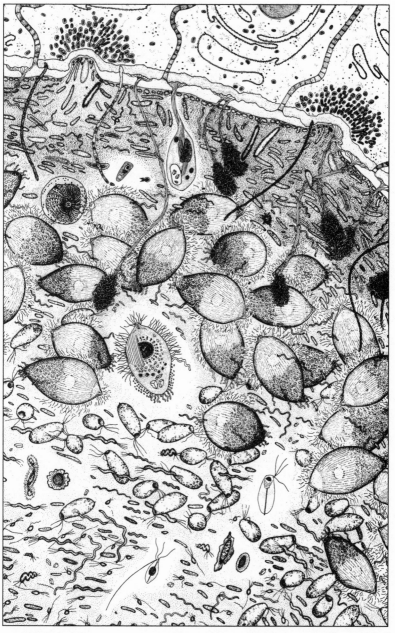

FIGURE 7.3 *Pterotermes occidentis,* Gut Symbionts of a Dry-Wood-Eating Termite

reside, hovers near ground level and occupies less than a third of the volume of the entire termitarium. The nest connects to the outside by radial tunnels, by vertical conduits, and ultimately by a central chimney closed at the top. Some nine and a half liters of oxygen flow through the *Macrotermes* city per hour, of which only one and a half are attributable to the termites. The other eight liters an hour can be traced to fungal respiration out in the indoor fields. The whole termitarium generates fifty-five watts of heat, of which roughly forty-seven are attributable to fungal metabolism and only eight to chemical transformation in termite bodies themselves.

Termites suppress fungal sexuality while they encourage growth of edible young "shoots." As long as these macrotermites prune and tend their gardens, the formation of mushrooms is entirely inhibited. Mushrooms are outward manifestations of the secret sex lives of fungi. Only when the termite farmers die do mushrooms ever sprout from termitaria. And the consequences of termites abandoning their caretaking responsibility are profound for people. Delicacies such as *Termitomyces titanicus*, a mushroom with a cap one hundred centimeters in diameter, appear on the abandoned termitaria. Schaechter calls termite mushrooms gathered from termite "ghost towns" "among the most prized in the markets of Nigeria, Zambia and tropical countries of southern Asia." Indeed it is only because termites occasionally leave their gardens fallow that we even know that *Termitomyces* fungi, of which apparently many kinds may be distinguished, even belong with Basidomycota, the mushroom fungi. The sexless fattened shoots, the staphyla growth form, is induced by the frenetic termite colonial behavior. This turgid, swollen-tip form is unique and different from the standard patterns of fungal growth. If the fungus-termite garden association were utterly permanent and termites were always in residence, we would be forced to classify the fungi as "imperfecti." Thousands of species of "fungi imperfecti" lack sexual stages entirely, for reasons we don't understand. By breeding spore farmers before sexual maturity, fungal expert Paul Stamets and his colleagues at Washington State found a way that he believes will rid houses of termites and carpenter ants. (He has patented it; see www.fungiperfecti.org on the Web.) The insects collect spores but when they mature, the fungal effect acts as a lethal all-natural population-level insecticide. Here we know the reason, at least superficially, of termites' more cultivating brethren: These termites entirely suppress the fungal sexuality. They keep the fungus in a constant nonsexual, food-producing mode. Human agriculturalists, interestingly, also suppress sexuality in some plant species. Branches may be grafted before they bear flowers or fruit or seedless varieties of fruits may be strongly selected.

The termite workers scurry among the crevices and cracks of the fungal city and its outlying properties as they forage for wood and other vegetation. Using their own capacious intestines as carrying cases, they enter termite city through many passages and channels. After they find a suitable garden plot they excrete the partially digested wood-fungal mash. At first, of course, this coarse material contains all manner of decaying fungi and other microbial gut dwellers. But somehow the worker termite's rear end becomes an inoculum . . . or at least an aseptic living trowel. The worker thumps down its partially digested mash on the floor of the inner chamber of the nest. "Once excreted, the fungal mycelium grows into tiny spheres, about the size of a small pinhead," wrote Schaechter. "These spheres, packed with fungal spores, are the most prominent feature of the fungus gardens. To the termites, the scene must appear as a field of tightly packed giant puffballs would to us." The abundant fungal food that sustains the city of two million workers and their charges is the pinhead spheres. The sedentary queen, the robust mandibled soldiers, and the newly hatched young termites must all be fed "proctodially" or "stomatodeally" by workers, that is, through anus-to-anus or mouth-to-mouth contact. The fungal spheres are hungrily grazed for breakfast, lunch, and dinner.

Termitaria are dynamic, spectacular high-rise cities built and maintained by hundreds of thousands, often millions, of insects in constant communication with each other. Birth chambers, hatcheries, the insect equivalents of schools, hospitals, honeymoon quarters, workshops, and morgues, supplied with gas and liquid fluxes, serve as climate-controlled insurance policies for the individuals they sustain. When marauded by an aardvark or battered by torrential summer storms, the citizens of southern African macrotermitine cities respond quickly to restore their first-class accommodations. Oxygen in water vapor-saturated air circulates briskly, carbon dioxide is kept between 2 percent and 5 percent. This CO_2 value is far higher than the 0.032 percent of the Earth's general atmosphere. In the dry season, the fungal gardens are kept moist by high internal humidity.

When their walled and chimneyed fortresses are threatened by potential disaster the termite citizenry respond predictably: bells toll, sirens screech, professional relief teams mobilize. A positive feedback strategy ensues called "stigmergy" (literally, *stigma* = sign, *ergy* = energy; thus they are "driven by the sign"). A termite stimulated at the site of the offense responds with an alarm reflex: She grabs a sand grain and cements it into place with a gluey mouth secretion. "Daubing," it is called. Then she releases a chemical alarm, a pheromone. This substance wafts through the channels and conduits of the city. Termites at the location of the disturbance tap their heads rapidly against the termitarium walls. The alarm at the site of the trouble quickly spreads from the bothered few to the multitudes. When other termites sense the residue of a fellow's daubing reflex, they daub too in response. Sympathetic daubing repeats. Pheromone wafting and head tapping travels from the site of the problem through the great pillars, halls, walled chambers, and gardens of the giant dwelling. This is "stigmergy." The group response is unambiguous: nestmates hasten to the troubled area. Within a few hours reconstruction is under way and the breach is sealed. Teams of worker termites erect new channels, galleries, columns, and halls. The whole architecture, argued J. Scott Turner in his wonderful book *The Extended Organism* (2000), is an "adaptive structure." Turner called the behavior that underlies the dynamism of Termite City "social homeostasis." The commotion's return to calm he described as follows: "A perturbation of the colony atmosphere elicits a response that returns the colony atmosphere to its state prior to the perturbation." The bacteria's, protists', fungi's, and termites' tendency to fuse to produce emergent structures is the same as ours, only they, in hundreds of species, urbanized at least 80 million years ago, and we have been at our fancy fusions for only some 8,000 years. The mergers and fusions that we feel are uniquely human in fact have many precedents in the nonhuman world. The termite farmers are merely a single example.

INDEPENDENT ORIGINS OF AGRICULTURE

Several times in Earth history, social animals have domesticated their food sources: Agriculture evolved in termites, ants, and people. "Domesticate" means "take into the house." Some agricultural associations became even more intimate: Food sources are not only grown very close to home but on the body itself. Here we tell one story of the origin of backyard farming and then briefly describe the even closer relationship of "body farming." The former association between termite and fungus sets the stage, while the latter, between bacteria and termite protist, demonstrates foreign genome acquisition. Food bacteria, food algae, and even food starch–storing ciliates are grown, respectively, inside the bodies of certain protists, green worms, and cows.

The lush tropical forest around the Tiputini Biodiversity Station in Ecuador can only be reached by a six-hour "canoa," a ride in a motorized unprotected river boat down the river from Coca. Termites, including the frenetic subterranean species *Heterotermes tenuis*, abound in the region, especially in fallen wood. Like all members of the family Rhinotermitidae, these masters of cellulose digestion harbor a varied and active crew of microbes in their bloated hindguts. On puncturing the gut one could recognize wood in the process of degeneration. Wood pieces can be seen directly through the translucent bodies of moving hypermastigote cells such as *Trichonympha* and some of the pyrsonymphids.

A few hundred members of the colony of busily bustling termites were brought into the laboratory but kept in their own home—a decaying log. Two weeks later, all over the decomposing branch sprouted what looked like hundreds of distinct, long-lasting drops of water, half a millimeter in diameter, but they weren't water. Microscopic observation showed the "water droplets" to be "sporodochia," a term we had to look up in a mycological dictionary. The sporodochia were packets of tissue of an obscure "fungus imperfectus" limited to the southern hemisphere. With help from some friends we located the extremely obscure and small scientific literature about them. The sporodochia made us wonder about fungal farming, so well known in the non-wood-eating "higher termites." Is "fungal farming" by wood-eating termites beginning anew in the tropics of South America? Are we witnessing an independent origin of agriculture?

Andrew Wier, a graduate student, electron microscopist, and scuba diver, returned from two weeks at the Tiputini Biodiversity Station in the winter of the first year of the new millennium. He came back to the Margulis laboratory full of the joys of the Ecuadorian rainforest. He seemed to recognize some of the great buttressed trees as *Iriartea*, members of the Aracaceae family. A profusion of pineapplelike bromeliads stick epiphytically on many of *Iriartea*'s branches and trunks. Certain bark-dwelling cyanobacteria turn trunks green, while white-paint-like fungi make them look like paper birds. The diversity of the luxuriant palms is marvelous. Wier was not too surprised to find that Tiputini was at too low an altitude to be home for *Gunnera manicata*, a huge symbiotic flowering plant that we will meet again in Chapter 11. He marveled at Bart Bouricious's Amazon treetop walkway, forty meters above the forest floor. Bouricious makes his living in an extremely uncrowded

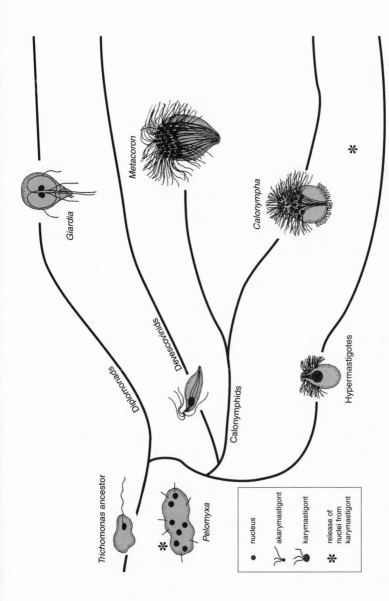

FIGURE 7.4   Archaeprotists: Archaemoebae, Diplomonads, Parabasalids (Devescovinids, Calonymphids)
and Hypermastigotes

field—he builds Amazon tree canopy walkways. Bouricious's access to the treetops has been displayed even in the temperate zones at Hampshire and Williams colleges in Massachusetts. To climb his ladder and observe macaws and spider monkeys on the horizontal walkway in the lush rainforest is an unforgettable experience.

As an accomplished field scientist, Wier spotted the actors—at least in general outline—on the Tiputini ecological stage. The recognition of higher taxa—palms, grasses, noisy cicadas, philodendrons, dumbcane, lichens, orchids, figs, conifers, ferns, polychete worms, isopod crustaceans, tree frogs, beetles, nematodes, bony fish—is of course not difficult. But general categorization is a far cry from the precise species identification needed for research. All Wier needed at first was the general knowledge to spot wet logs loaded with wood-eating termites. At the time he visited Ecuador he was studying an organism called *Staurojoenina*. A termite protist of gorgeous complexity, *Staurojoenina* avoids oxygen. This single-celled swimmer, called a "nymph" (or hypermastigote), farms its own bacterial body parts. Wier had discovered large numbers of this magnificent wood-eating nymph swimming inside the puffed-up intestine of a wood-dwelling termite, tentatively identified as *Neotermes mona*, that his fellow naturalist Sean Werle brought back from Trinidad. Wier explored the termite nymph as a part of his master's degree research on a fungus called *Cryphonectria*. This whitish fuzz fungus is associated with the dieback of mangrove swamp trees near La Parguera, the biology station of the University of Mayaguez, and at the coast of Phosphorescent Bay in Puerto Rico.

The world of termites, fungi, and mangrove trees had become Wier's world on his four research trips to Puerto Rico. Just before he left for Ecuador he decided that "his" new *Staurojoenina* protist was the same species as one well established in the termite literature called *Staurojoenina assimilis* (Figure 7.5). Using a scanning electron microscope, Wier found that the protist was covered with a great investment of rod-shaped bacteria, scores of tightly packed epibionts. These rod bacteria extended all over the nymph's anterior surface. Wier concluded they were probably the same bacteria as those revealed in the spectacular transmission electron micrographs of *Staurojoenina* made by David Chase in the early 1980s. He sent a few of Werle's dead termite soldiers in a vial filled with alcohol to an expert, Rudi Scheffrahn at the University of Miami Research and Education Center, for identification. Scheffrahn is a whiz at the recognition of "lower" (wood-eating) termites—even though there are some estimated 400 species of live ones and even more that are extinct. Back in the lab we concluded that we had *Heterotermes tenuis*, a common lower termite from the new world tropics.

Within a week the tiny translucent dots appeared in irregular arrays all over the rotten wood in which the termites were living. By two or three weeks the dots reached pinhead size, and so they remained for months. This colony of *H. tenuis* didn't mind a good dousing with "rain" (distilled water)—in this it was unlike any other lower termite in our experience. Overwatering inevitably causes termites to succumb to fungus takeover and die.

When we examined the translucent, gummy "raindrops" we immediately saw that they were closer to a pure culture of fungal spores than any of us had ever seen before in a natural sample. Furthermore, the spores were large three-celled affairs that were not compact and travel-ready. They were swollen and turgid, more suggestive of food than propagules. In the absence of the sex organs of molds and mushrooms, of asci and basidia, we assumed the fungi were "imperfecti"—that is, deuteromycota, but we failed to find this large, conspicuous form of spores in any of our reference books. Nor could our colleagues help us. Only after we sent our information and observations to the superb mycologist Kris Pirozynski in Ottawa did we receive the crucial lead. "I recognize those spores," Pirozynski said. "They are *Delortia palmicola*, the same fungus I tallied in my mycological census in Kenya nearly sixty years ago." Then he asked us if the wood they were found in could possibly be

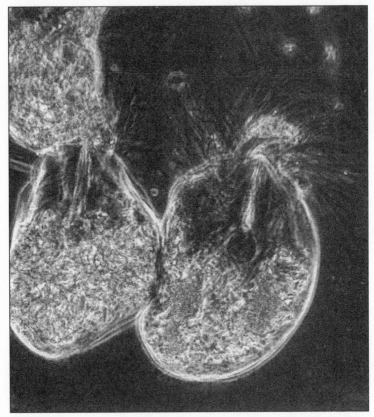

FIGURE 7.5   *Staurojoenina:* Composite Individual Hypermastigote

from a palm tree. He explained that the fungus was discovered by Monsieur N. Patouillard in French Guiana. Patouillard published descriptions of fungi on palms, including drawings, which exactly matched what we observed—in 1888. As far as Pirozynski knew, this *Delortia* was entirely restricted to palm wood from the southern hemisphere. He figured it had evolved on the great Paleozoic southern continent of Gondwanaland and never spread north to Laurasia. Pirozynski had seen dried and shriveled sporodochia more than once on felled palm trees. He always suspected they were associated with some insect but he had never seen any insect on the shriveled remains of the fungus. When we put our observations together we concluded that the pinhead "permanent raindrops" fit the definition of sporodochia. The badly decayed wood was likely to be palm because the Tiputini collection site was loaded with palm trees.

Evidence accrued when one of us (Lynn Margulis) was fortunate enough to visit the Tiputini Biodiversity Station in early June of 2001 as a guest of the Universidad de San Francisco Quito, Ecuador. With the help of a native guide, Mayer Rodriguez, and microbiology professor Gabriel Trueba, she revisited Wier's *Heterotermes* in their native habitat of soaked wood. The felled palm *Asterocaryum chambira*, a majestic thirty-five to forty meters tall when fully grown, supplied the food and nesting materials to the busy wet colonies of wood-eating subterranean termites. They were happily at work munching on the palm bits when we sacrificed some workers to examine their innards. The presence of a trichonymphid, a spirotrichonymphid, and some pyrsonymphid protists was enough to reassure us that Wier's *Heterotermes tenuis* either had been growing on palmwood—or that they could have been. These were clearly wood-eating "lower" termites, living under the wettest conditions we had ever seen termites endure. Locals say the Napo River and its Tiputini tributary have two seasons a year: *lluviosa y lluviosa* (rainy and rainy).

Pirozynski reasoned that in Africa the fungus *Delortia palmicola* probably grew well on damp, especially dying, palmwood. There too the *Delortia* sporodochia probably also fed termites even though they did it out of Pirozynski's sight. All the observations now were consistent. They jibed with another discovery of ours: the abundance of *Delortia* spores in the intestines of our *Heterotermes*. Along with wood-digesting nymphs and myriad bacteria, their hindguts were filled with the unmistakable three-celled *Delortia palmicola* spores.

Care in watering, in our long experience, is crucial to the maintenance of termite colonies in the laboratory. Excessive moisture always encourages fungal growth in boxed-in isopterans. Just a single mistake by an enthusiastic undergraduate who overwaters, who fails to notice any standing water, kills an entire colony in a weekend. We were therefore amazed that *H. tenuis* subcolonies flooded with "rain" continued to thrive. They apparently respond by a helter-skelter business-as-usual recovery that, to our untrained eyes, seemed comparable to J. Scott Turner's description of "stigmergy." The agitated termites waved their antennae and, it looked to us, during the heavily watered weekend patched the collapsed portions of their abused, still-damp quarters. Flood waters and heavy rains, we surmised, must be business as usual in their Ecuadorian habitat.

Prototaxis applies to all of the complex relationships between fungi, termite, protist, and wood. Relationships begin casually, they become intimate, and new, unique individuals evolve.

Might not the adage that "the best defense is a good offense" have been the first stage of strategy? Fungi and rain must have been constant threats to *Heterotermes tenuis*'s rainforest ancestors. But those termite colonies that coped with the threat of standing water and began to eat the fungal intruders thrived. Presumably the *Delortia* fungus harbors at least some cellulase enzymes it can put to use for wood breakdown. When a *Heterotermes tenuis* termite eats *Delortia* it gets two for the price of one: consumable nitrogen, from fungal cell-wall chitin and protein, and cellulase enzymes from the live fungi on loan.

Today's *Heterotermes tenuis* in the Tiputini wilderness may or may not be on the road to becoming a "higher" termite. All were still replete with wood-digesting nymphs when we looked. But perhaps among these Ecuadorians dwells a new fungal farmer, devoid of cellulolytic protists, ready to take up urban architecture. Perhaps not. These water-loving termites seem remarkably to resemble their Mesozoic ancestors, which did eventually choose fungal farming over cultivating microbes in their intestines. For awhile the ancestors probably supported themselves by both modes. This possible evolutionary pathway from lower to higher involves both speciation and symbiogenesis. Before the higher termites farmed fungi in huge, environmentally controlled termitaria, certain protist-infested lower termites defended themselves from fungal onslaught by domestication and ingestion of their offenders. The termites learned the trick of "pure culture" fungal growth as they prompted the formation of water-dropletlike sporodochia in massive quantities on the surface of their resident logs. They pruned, planted, and ate spores in these "kitchen gardens" much as our hunter-gatherer ancestors took the first steps toward agriculture by tending (and defecating near, thus fertilizing) naturally occurring patches of edible grass. Their fungal culture techniques, we infer, preceded mound-building. The cultivation occurred on fallen logs in dark rainforests and wet fields before this line of termites constructed their fungi-forming cities.

Termite dependence on fungal farming passed the point of no return when the insects lost their wood-digesting protists. Cellulases, enzymes that degrade wood, are rare or absent in termite bacteria. Termites cannot survive solely on wood dissolution by bacteria alone. They lost protist wood digestion, we suspect, after sporodochial fungi like *Delortia palmicola* were tamed. Domestication occurred in the southern hemisphere, most likely, because this is where large-scale gardening by higher termites seems to survive. It is because fungal foes became fungal allies that we can contemplate landscapes like the one shown in Figure 7.6, photographed in the Australian outback by Reg Morrison. The successful real estate company of these macrotermitids ultimately is owed to their being mycological maestros. One huge group of cellulose-digesting symbionts, the amitochondriate protists, was replaced by another successful group of cellulose-digesting microbes, the fungi. Fungi are more modern than protist nymphs—they thrive in the open air, resist desiccation, and

FIGURE 7.6   Termite City

stay where they are put. Like our corn, wheat, and barley, they are
amenable to domestication—to being taken "into the house."

In studies of termites—a world inside a world—no bacteria digest wood. No termite, bereft of its protists, does either. The solution
to the fierce selection pressure to extract food from the environment,
especially from refractory solid wood, led to farming. One set of termites abandoned wood; although abundant, its nutrition was too inaccessible. Another set grew a menagerie of swimming protists inside
their guts, each protist covered or filled with entirely distinguishable
bacteria genomes—entire rod, or spherical, or filamentous bacteria
(or all three types) with their genomes intact. In these protists, who
never indulge in sex, the acquisition and integration of permanent
bacterial symbionts is what led to new species.

PART THREE

PLANETARY
LEGACY

CHAPTER 8

GAIAN PLANET

*I*n the growth of modern cities from ancestral human tribes, many trades and specialties have appeared. The general trend in the 100,000-year history of human migration out of central Africa has been for small bands of hunters and plant gatherers to settle into local population centers. Seasonal migration patterns were replaced with permanent gardens and expanded fields of domesticated plants and animals. By 11,000 years ago the major sustaining economic forms of life had co-evolved with the upright prattling ex-African; dogs, cats, and grasses had been pressed into service by extended families, bands, and tribes. Gardens of root vegetables, edible flowers, and leaves were cultivated in the tropics and subtropics worldwide. In temperate Europe and Asia, sheep, cattle, and chickens helped support a great rise in human settlement.

In the days of the hunt, which in many places lasted from earliest human times through the nineteenth century, ancestral males, mostly, ran for days to track game. The effective loss of heat as sweat was made easier by loss of body hair. Men hunted in communicating groups while women were more likely to gather plants and tend children. Perhaps select females ("ancestral pharmacists") became expert at the recognition of plants with medicinal properties. At certain times and places, certain males, often father-to-son lineages, were trained as "fathers" of the tribe. We understand them in retrospect as the ancestral politician-priest-shaman rolled into one. Today the social roles proliferate as more and more specialties come into being. We have butchers, bakers, candle-makers, baseball players, auto mechanics, insurance agents, and television repairmen. Our schools and universities record a huge diversity of academic "fields." The biological sciences, for example, are divided into great subspecialties: anatomy and ultrastructure, microbiology, cell biology, paleontology, physiology, genetics, molecular biology, and so on. But of course they all study aspects of only one living Earth. Academic territoriality, like the emergence of many trades in modern society, reflects the inexorable growth of human populations and the increased efficiencies gained from division of labor. Specialization is part of the answer to how philosophy grew into modern science: Some theorize and others observe, and the observed areas become smaller each decade as the number of data gatherers increases.

This trend toward increasing specialization is not limited to humans. Specialization increases in ecosystems as they develop. When forest or riverbank ecosystems grow back after a fire, flood, or other perturbation, recovery begins with fast-growing populations of organisms of the same kind. Such pioneers (certain bacteria, algae, grasses depending on the particular system) all behave in much the same way. Their populations grow rapidly to fill the available niche-space. But then, inevitably, they face environmental limits that restrict further growth. Lack of space, water, phosphorus, or other nutrients blocks further population expansion. At this time more slowly growing species, which were unable to settle the original environment, join the fast-growers to make more efficient use of nutrients, energy sources, water, or other limited environmental variables. The slowest-growing and most stable communities of organisms at a given location, the so-called climax ecosystem, ultimately replace the pioneers, sometimes after a long succession of intermediate stages. These climax groups of plants, animals, and microbes tend to have the greatest diversity of species, the most complex interactions, and the highest energy efficiencies relative to their predecessor communities.

Specialization, whether of ecosystems, scientific investigators, or modern urban trades, generates organization within populations of huge numbers of individuals. Consider an example close to our home. One of the densest human populations on Earth is in our small, bucolic town of Amherst, Massachusetts. The density of the student population in the Southwest Residential Area at the University of Massachusetts is 5,400 in one-quarter square mile. New York City only boasts 1,500 residents in the same amount of space. The reason is simple: Cities like New York are densely populated because of skyscraper office and apartment buildings, many empty each night. Southwest residential dormitories, the tallest only twenty-two stories high, support a vast number of human lives per square foot of floor space because the students live in packed rows. The two students per room lack kitchens, private bathrooms, and open space. New Yorkers enjoy many parks, notably Central Park, and usually more than just one-half bedroom per person. In the U-Mass high-rise dormitories, specialized buildings devoted to food service supply hundreds of students in dedicated dining halls. Open space, athletic fields, and common rooms are not close by. Manhattan dwellings tend to have bathroom, kitchen, and living areas in each unit, and are not specialized by building.

The efficiencies of specialization bring mixed results. Although effective for organization of numerous individuals, specialization separates us from our sense of the whole. Science, as an international activity, may be more effective now than in the past, but Renaissance polymaths familiar with virtually all of knowledge are increasingly scarce, even nonexistent. No one anymore can be familiar with all branches of culture. Yet it is perhaps possible to see the whole. This may be one reason for the growth of popular science, which allows the layperson to see the whole—at least from the viewpoint of the natural sciences—and forces the scientist or science writer to describe such a whole. It is also the reason for the appeal of evolutionary science. Evolution is a science of connection.

Before evolutionary science made its inroads into human understanding, the operating notion of the whole was most often a religious one: the single Christian, Jewish, and Moslem God. Each kind of life, each named species thought to have been separately created by God, reflected a variation on the divine design. The human interpreter of that design, whether priest or saint, occupied a privileged spot. But Lamarck and Darwin envisioned a more realistic wholeness. They outlined an understanding of how all species of organisms are linked not by an all-creating Deity, but by time. The genius of Darwin, and the many other natural historians, breeders, and geologists whose works he and his successors systematized, was to show that all organisms alive today are connected in time by common ancestry.

Most educated people believe they understand the full implications of this connecting insight. We suggest that certain basic implications of Darwin's evolution have not been grasped. We see connection between the growth of an animal from a fertilized egg and that of the biosphere from the first cells. Both entities develop, they evolve by change through time. Both growing children and biospheric evolution are akin to the expansion of diverse ecosystems. Change through time of live beings requires an increase in number of cells—and accompanying cell death. Although many educated people seem comfortable with the notion that we evolved from species kindred to modern apes, the idea that we are ultimately descended from bacteria—germs—is seldom emphasized. That our own bodies are hugely overpopulated cell clusters, the natural clones of living and incessantly dying cells with nuclei, is hardly ever stressed. But we come from symbiotic microbes, from nucle-ated specialized cells, from wormlike and lungfishlike animals, from reptiles and insectivores, prosimians and anthropoid apes—from form upon form. At no time in the past since the first life began did our lineage or the system itself ever die.

VERNADSKY'S SPACE, DARWIN'S TIME

Connecting ourselves to the full panorama of Earth history requires a vision extending beyond our immediate ape antecedents and even beyond the mammals and reptiles that preceded them. Microbes were the first life, and changes in microbial communities were the basis of the first evolution. The transformation of crowds of cells into individual larger organisms, although omitted from classical zoocentric evolutionary theory, is part of Darwin's legacy. Darwin's legacy is that of the connectedness of life through time.

Vernadsky's legacy is the connection of life through space.

Vladimir I. Vernadsky (1863–1945) was a remarkable individ-ual whose contribution to the sciences—including the extension of biology to other sciences—is far more appreciated in the east than the west. In the former Soviet Union some fourteen institutes are named after him. Air letters and postage stamps carry his bearded visage. Perhaps the official atheism of the Soviet Union created a scientific and intellectual climate where the ideas of Vernadsky were easily assimilated. He was years ahead of his time in his view that life is the most important geological force and the Earth is a single interconnected life-supporting entity. Although Swiss geologist Edouard Suess coined the term "biosphere" (literally, sphere of life) to complement "hydrosphere" (Earth's layer of water) and "atmos-phere," Vernadsky via his 1926 book *The Biosphere* brought the term into common use. This major work was translated into French in 1929, but the whole book was not available in English until 1998. Vernadsky, a chemist who studied crystallography, was always interested in the structure of minerals. He was a scientific monist with novel insights. Unlike the vitalists, who held that there must be a special, unique property of life that gives it the ability to think and act on its own, Vernadsky saw life as a natural outgrowth of a chemically active universe. The early vitalists argued that life must be made of special stuff that distinguishes it from nonliving matter. But after Wohler synthesized urea, chemically related to the uric acid of urine—a chemical containing nitrogen, oxygen, and carbon—in his laboratory in 1828, the argument that life is made of special matter would not hold. Vernadsky took the notion a step further: He looked at life as an entirely natural global phenomenon. Indeed the very word "life" contains a vitalistic connotation, that of a special entity that cannot be reduced to physics or chemistry. To avoid this connotation, Vernadsky resolutely used the term "living matter" instead of "life."

Influenced by the movement of arms, men, tanks, and planes on a global scale in World War I, Vernadsky saw life (including human life) as a complex mineral. Living material was a kind of "animated water." A multicomponent fluid, a kind of solid, life was a highly charged, energy-rich, and reactive form of matter. With chemical impurities and in a special phase, to be sure, life is still mostly salty water. Over evolutionary time, Vernadsky argued, more and more chemical elements became involved in the global circulation of living matter, and the rates of elemental circulation—of carbon, hydrogen, nitrogen, oxygen, sulfur, and phosphorus—tended to increase. The complex mineral that is life is also horizontally transportable (for example, by wind or water) or self-transporting. Locusts, for instance, descend upon and devour fields of grain, then swarm again to the skies to transport organic-rich salty water horizontally over the Earth. Such massive, moving populations of insects were, for Vernadsky, "flying mountains."

Living matter, according to Vernadsky and his modern spokesman Andre Lapo (1987), is the strongest geological force. The active part of the biosphere, a thin layer of life twenty to thirty kilometers deep, ranges from microbial spores in the atmosphere to ecosystems at the ocean abyss. The active part of Earth's surface resembles the bark of a growing tree. The wood of the tree is not so much dead as composed of formerly living tissue integrated into the still-functioning water-sugar transport system of the heartwood. Thus much of the landscape we see as mineral—sand, rubble, limestone, iron ore—was shaped by activities of life or went through a phase of production inside cells. They trace "bygone biospheres." The white cliffs of Dover are fossilized skeleton-forming microbes, as is most limestone worldwide. Soil, one of Vernadsky's earliest objects of interest, would not exist without intense biological activity. Soil also interested Darwin, who wrote a monograph on how earthworms help produce soil.

Soil does not exist on Mars or Venus; rubble called "regolith" does. Moist, fertile, and replete with protist cells and other organisms, soils are a kind of tissue of a living Earth. Their composition is determined by the worldwide circulation of key elements. Because no individual organism is immortal and the materials of which all are composed are limited, crucial chemical elements (carbon, hydrogen, oxygen, sulfur, phosphorus, and nitrogen) recycle in all ecosystems and on a global scale. The computer models of geochemists show how carbon, oxygen, nitrogen, water, and other molecules circulate at the planet's surface. Accurate model-making has become a scientific goal among geochemists and biochemists in recent years after Vernadsky and Lapo focused interest on Earth's life as deeply intertwined with geology. The trend toward linking biology and geology has become strong, especially since James Lovelock's publication of his highly original work. The brilliant introduction of the "Gaia Hypothesis" was a wake-up call to the scientific community to consider "life and the environment as a tightly coupled single system," as Lovelock put it. Entirely independently of the Vernadsky legacy, Lovelock, a chemist, measured astounding chemical discrepancies in the Earth's atmosphere. He linked these anomalies to life's activities. Since the late 1950s Lovelock has refined, expanded, and improved descriptions of our lively, unique, and watery planet. Increasingly, model makers have become aware of what academic specialization tends to obscure: Life—part of the "environment"—cannot be ignored in any of these models. Matter circulates in the biosphere just as it does inside our bodies. Both the biosphere and the individual vertebrate animal contain mineral hard parts—bones, teeth, scales, hair—that are not, in and of themselves, alive.

The analogy between mammalian bones, trees, and the living Earth extends deeper than you might think. Whereas life cannot claim an influence over Earth's molten core, it probably plays a prominent role in such apparently purely geological phenomena as the presence of water on Earth (kept here because bacteria, plants, and algae remove greenhouse gases from the air), weathering (because bacteria and fungi pulverize and digest rock, subjecting it to erosion), and even plate tectonics (as discussed below).

One great advance since Vernadsky's recognition of life as a geological force has been James Lovelock's Gaia Hypothesis. Lovelock recognizes that Earth's atmospheric chemistry, its mean global temperature, and its oceans' salinity and the alkalinity (pH 8.2) of its surface environments are not random. They are regulated, presumably by the metabolism of the sum of Earth life. This sort of global modulation does not mean the Earth's surface is equivalent to an organism (which cannot, like the biosphere, survive on its own waste and breathe its own gas excretions). Yet the Earth's surface certainly has organism-like traits. It is built largely of reproducing cells, takes in nutrients and water, and incessantly produces wastes. Each comes into ecological, sometimes symbiotic associations that become absolutely required—to recycle wastes. This expands the cellular realm. The result is that, over time, the environment becomes increasingly organized, differentiated, specialized.

FIGURE 8.1   The *Apollo* Astronaut's Earth, as Seen from the Moon.

Parts of the environment may ultimately become organized into the body or an extension of the body. We do not think of the calcium in our bones as "environment," but that is exactly how bones and many other hard parts began in the history of nucleated cells. Calcium, a toxic waste, overabundant in seawater, had to be removed. It was metabolized into calcium carbonate ($CaCO_2$), or calcium phosphate, hard substances that accumulated in cells, which were then used for protection and structural support. Among the colonies of cells for which these hard parts proved useful were our fishy ancestors. Others included the ancestors of mollusks, arthropods, and the hard-shelled plankton known as foraminifera and coccolithophorids that, falling in a constant rain over millions of years, created the chalky seabed that when uplifted, ultimately became the white cliffs of Dover (for more, see T. H. Huxley's famous essay "On a Piece of Chalk"). This process had two effects: It created a major landform that is common worldwide (chalk or limestone sediments), and by burying large quantities of calcium and carbon underground, it reduced the levels of these elements in the atmosphere and ocean.

The process of incorporation or reutilization of wastes, of growth and recycling, of economical use of limited materials in a closed space, links the Gaian notion that Earth has a physiology with Vernadsky's emphasis on living matter as an extension of geology. Whereas Vernadsky deconstructed the hierarchy between earthly life and nonlife by demonstrating the mineral nature of living matter, Lovelock did the same by showing several physiological behaviors of the entire biosphere. Both men's views reflect the spatial interconnection of the thermodynamically open systems we call living matter with the environment from which it derives its energy. The chemistry of our atmosphere is not at all random. It directly connects to the breath—the intake and output of gases—by many trillions of cells. Each cell, each body, each city interchanges gases with Earth's atmosphere, waters, and soils all the time.

The geological revolution of the 1960s and 1970s established without a doubt that the Earth's surface is covered with huge, wet, mobile lithospheric plates. Prior to the establishment of plate tectonic theory as explanation for continental drift the obvious was assumed to be true: The Earth is solid under our feet, the ground now is where it has always been—and, except for earthquakes, the land and sea are unchanged in their distribution. Primarily composed of heavier basalt and lighter granite, these enormous plates of solid rock, we now know, are produced at receding boundaries and destroyed at violently colliding or subducting boundaries.

The tendency of the continental plates to move laterally on the Earth's molten mantle may itself be connected to life. The presence of liquid water on Earth's surface for hundreds of millions of years has usually been attributed to our planet's lucky position. But Earth's water might have dissipated to space long ago, as it apparently did on both our neighbors, Mars and Venus. In the absence of the recycling activities of the living and the removal by the living of carbon dioxide from the atmosphere, the Earth too might have lost water. Metabolism, growth, and reproduction of cells retained it. Photosynthesis removed the greenhouse gas carbon dioxide from the atmosphere to produce all manner of organic compounds, and thus kept the planet cool enough that its water vapor did not escape into space. Without water for lubrication the movement of crustal plates might have ground to a halt. The wet living Earth's surface retained environmental conditions characteristic of an earlier solar system. Life, as Vernadsky sketched and Lovelock painted it, is a planetary phenomenon.

We can now begin to see how Vernadskian space and Darwinian time are themselves connected to the rest of the physical universe. Through metabolism and reproduction, Earth's life tends to maintain thermodynamic environments characteristic of earlier stages of the planet's development. Our own watery organic bodies, like those of all animals, plants, and microbes, are a kind of time capsule that contains Earth's chemical environment as it was three billion years ago. The ancient past is preserved within by prodigious fluxes of solar energy captured by photosynthesis or deep hot chemical reactions in the bowels of the Earth. The environment within which evolution occurs is dynamically stable and self-regulating. Largely maintained by the chemical and biological interactions of members of microbial communities, the ocean-blue biosphere's physiological stability requires an incessant flux of energy from outside the system. Life is Earth-bound and cellular, but also a geological and solar phenomenon.

ARCHEAN, PROTEROZOIC, PHANEROZOIC AGES OF GAIA

When James Lovelock named his hypothesis "Gaia," it was just another of his strokes of genius. The name (from the ancient Greek

Earth goddess), suggested by his Wiltshire neighbor, the novelist William Golding, was far catchier than "Earth as a homeostatic chemical system." The Earth is a physiological system, he argued, and as in any physiological system, regulation can be detected: temperature, gas flux and composition, salt, acidity-alkalinity concentrations. They are dynamically stable. Physiology is easily inferred from measurements. By giving the planet the name of a Greek goddess, Lovelock, in one act of baptism, made his idea memorable to both science and its enemies. He personalized the object of study and focused the attention of his colleagues on her salient features. "Gaia" was reduced, in the minds of many, to a slogan: "The Earth is a giant organism"—indeed a giant "female" organism. How, we complained to Lovelock, can the Earth be "an organism" when no organism lives off its own waste? Why do you persist in promoting the image of the Earth as a giant single being, we asked, when it detracts from the science and infuriates potential serious colleagues? Why not call Gaia a giant ecosystem, which recognizes the plurality of its component beings, we suggested. Lovelock disagrees. He finds the term "ecosystem" cumbersome, difficult to define, and entirely opaque to those unfamiliar with ecology. Furthermore, he defends his metaphor of the goddess. "If they think the Earth is a live being," he explains, "they respect and love Her. If the planet is only a pile of rocks it accepts kicks, jolts, and hard knocks. No one cares about it. Of course Gaia is not a single organism, nor any sort of goddess, but she is alive and deserves our reverence and understanding."

Lovelock's best book, *The Ages of Gaia* (1988), engagingly explains the development of the Gaia Hypothesis, now the Gaia Theory, as well as the development of the planet itself. From the birth of the solar system to which it belongs, through adolescence and adulthood, we trace the Earth as living system. We are privy to the science and scientists who show us how its most personal secrets were divulged. We are accorded Gaia's 4,000-million-year medical history, track her many changes and maturation steps, and even glimpse her likely old age and death. Lovelock's analogy between our own human developmental sequences and their predictable sensibilities and the planet-sized regulatory system he calls Gaia's physiology actually works. Although "geophysiology" sounds more scientific, Gaia stuck. "Gaia" it is, and "Gaia" it will stay. Gaia is the interactive system on the surface of the Earth, supplied with solar and geothermal energy gradients, that maintains the temperature, close to 18 degrees Centigrade. In the face of acid threats its alkalinity is pH 8.2. The atmospheric pressure of reactive gas (oxygen, for example, at 21 percent) is far from equilibrium. New Gaian science, much of it masquerading under the name ESS (*E*arth *S*ystem *S*cience) is robust. For more, see "Gaia and the Colonization of Mars," "Gaia and Philosophy," and "A Good Four-Letter Word," in Part III Gaia (pages 125–261) in *Slanted Truths* (1997) where we explain the Gaia Hypothesis and its implications for our lives.

The original Gaia theory involved three actively maintained systems: temperature, chemical composition of the reactive gases (oxygen in particular), and acidity-alkalinity. New work will try to answer further questions generated by Lovelock's Gaia: These include "Water Gaia," or "Has water been retained on Earth (relative to Mars and Venus) because of life?," the size of Gaia ("How deep does life and its attendant organic compounds extend into the Earth's interior?"), and "Is salt concentration, or that of other minerals, a Gaian phenomenon?" Others spring to mind: Is granite a Gaian rock? Is the distribution of great iron formations in time and space related directly to the genesis and development of life? As Lovelock more than once pointed out, the power of a good scientific theory is less what it definitively explains and more about what questions, good observations, and precise experiments it stimulates. On this criterion, Gaia is a scientific theory. The personification of Gaia in the "Ages" book is not a publishing gimmick; it is an aid to disseminate the basic tenets of the complex idea.

The metaphor of Gaia as an ancient strong-willed goddess allowed Lovelock to write about Earth's history in a fashion accessible

to readers. Some readers may be unfamiliar with "time-rock division," the great deep-time scale that organizes immense quantities
of geological data. A Gaian view of deep time is useful here. "Eons"
are the largest divisions of the geological scale, and after the Hadean
(hellish) eon before any life existed, there are only three: the
Archean, Proterozoic, and Phanerozoic Eons. We need to be ac-
quainted with all three to understand the origins of species. Gaia,
the living system, was born at the very beginning of the Archean
Eon, about 4,000 million years ago. The Archean was an age of
great tectonic and meteoritic activity, when the Earth's crust solidi-
fied from molten rock and the release of gases from the planet's interior created the atmosphere. This eon lasted for 1,500 million
years. Most of the rocks of the Archean Eon are dark basalts and its
organisms are the bacteria.

During the Archean eon the Earth had little oxygen in its rocks,
air, and water. A great buildup of oxygen began some 2,500 million
years ago, marking the beginning of the Proterozoic eon. By 1,800
million years ago atmospheric oxygen was plentiful, granitic rocks
dominated continents, and the nucleated cell had appeared. Biologically, the Proterozoic eon is the age of the aerobic bacteria and the
protoctists. These two first Gaian eons, the eons of the microcosm,
ended 541 millions years ago at the base of the Cambrian. The
Phanerozoic, the shortest eon and the most familiar, begins at the
Precambrian/Cambrian boundary. This boundary is now officially
marked at Mistaken Point, on the Avalon peninsula, Newfound-
land, Canada. This location where Cambrian (Phanerozoic Eon)
fossils are underlain by a plethora of Proterozoic Eon fossils, the
Ediacaran biota, is relatively accessible. An even more geologically
dramatic fossiliferous locality such as the Aldan River in Siberia is
inaccessible for those who lack time, money, large boats, and technical climbing skills. In many other Precambrian-Cambrian border
rock outcrops, preservation is poor, sedimentary sequences are missing, or access is nearly impossible. The present eon began with the

famous explosion of animal hard parts: sclerites, shells, trilobite carapaces, calcium-carbonate covered algae, and fecal pellets. The earliest Phanerozoic eon is still misconstrued in museums around the world as the dawn of life. The well-known divisions of the Phanerozoic include the Paleozoic, the age of shelly marine animals; the Mesozoic, the age of reptiles; and the Cenozoic, the age of mammals. But Gaia was already a mature matron when the Phanerozoic began. Some 450 million years ago she gained her green coat: Forest trees and low-lying plants began to decorate land surfaces once coated only by cyanobacteria and soil algae.

By the time the Phanerozoic eon comes into view at 541 million years ago, Gaia shows many signs of age. She has already suffered the greatest threat to her survival since her tumultuous birth in the lower Archean. The event was far, far worse than the great loss of the dinosaurs when many of their planet mates were snuffed after the enormous Chibchulub meteorite hit the Earth 65 million years ago and changed the climate. Even the loss of 75 percent of the largest taxa 225 million years ago at the so-called P/T boundary—the Permo-Triassic extinction—pales by comparison. Rather Gaia's life, the entire planetary physiological system, was threatened with death when the greatest ice age of all set in: the "Snowball Earth" of some 600 million years ago, for which there is late Proterozoic evidence around the world. Glaciers penetrated even the equatorial regions. Carbonate rock accumulated in huge layers, carbon dioxide built up under the ice, and eventually, as reduced photosynthesis led to an accumulation of greenhouse gases, the atmosphere heated up and the ice melted. This cycle recurred several times before Gaia ultimately prevailed.

Within a hundred million years, by 541 million years ago, the Earth was teeming with larger, more diverse, and more stunning forms of life than before. The great English-American ecologist G. Evelyn Hutchinson described Gaia's story as the ecological play in the evolutionary theater. The theater, the Gaian planet, persists but the actors in the ecological drama change. Some quit, others are fired. New species form by combinations and alliances among the old. Random mutations and other genetic changes hone and refine the new allegiances. The ecological play on the evolutionary stage goes on with new young actors and recast old ones, incessantly changing roles, and scenes by new playwrights and poets. Gaia's action is worth watching and worth your participation. It will perhaps continue for another 4,000 million years. Gaia will then die because the sun itself will die. Until then the drama will continue to unfold and Gaia herself, with or without humans, will probably continue to thrive.

CHAPTER 9

EUKARYOSIS IN
AN ANOXIC WORLD

"*E*ukaryosis" is the transition from bacteria to our kind of (that is, eukaryotic) cells. In the history of life cells with nuclei, eukaryotic cells, first appear in the fossil record during the Proterozoic eon. The earliest evolution of living beings made of single nucleated cells, protists, left descendants: amebas, euglenas, ciliates like euplotes or paramecium. All visible organisms—plants, animals, and the "large microbes"—are composed of many such cells. Put simply, eukaryosis refers to the largest evolutionary step that ever occurred in life's checkered history. No missing links between eukaryotes and bacteria exist, either in the fossils or in life. The sudden appearance of eukaryotes on the evolutionary stage was genuinely discontinuous and not gradual. How did it happen? How is the origin of nucleated cells related to us, to people?

Most people unselfconsciously subscribe to a certain folk classification. The world's life is divided into three great categories: Animals,

Plants, and Germs. (Other people assume at least tacitly four categories: Animals, Plants, Germs, and People; but those who deny the animal nature of people are unlikely to be reading this book.) Every educated person supports medical research that assumes humans are mammals akin to rats and mice. Anyone who looks into it agrees that people have bony skeletons with tail rudiments, opposable thumbs, three-dimensional vision, teeth, and other features that place us squarely among the primate mammals. Even with scant scientific knowledge, the thoughtful reader can recognize instantly the chordate (backbone, nervous system with brain), mammalian (hair, mammary glands) and primate (chimplike hands, teeth, posture, and social behavior) nature of all people. Of course all people today belong to the same species, *Homo sapiens*—because all, in principle, may mate and produce potentially fertile offspring.

Microbes (bacteria, the smaller protoctists, and the smaller fungi) were first discovered by the Dutch draper Antony van Leeuwenhoek (1632–1723), who made his own microscopes. The French chemist Louis Pasteur revealed a microbial world very different from the ordinary one of large organisms. Pasteur's practices mightily influenced a new science where the relationship of bacteria and yeast to people was studied intensively. Was the presence of the growing little spots (populations of bacteria) correlated with childbed fever? Did a virus cause foot-and-mouth disease? Did the apparent ability of the little budding spheres to produce alcohol correlate with the tendency of grape juice to become wine? Even today basic knowledge about the lives of microbes is hard to find in the medical, biological, and nutrition literatures. "Facts" are taught on a need-to-know basis. *Spirulina*, for example, is a common health food additive. Although it is unambiguously an oxygen-producing photosynthetic bacterium, it is marketed as an "alga." Honesty is not always the best policy: To deliberately promote drinking a bacterial solution could be commercial folly.

The similarity of plants and animals is not intuitively obvious, but they are far more closely related to each other than either is to bacteria. A far larger discontinuity exists in nature between bacteria and all others than between any plant and any animal. This concept is difficult even for naturalists and others who know the microbial world exists, but since the end of the nineteenth century, work by hundreds of scientists has been converging on this relatively new idea. The largest evolutionary discontinuity on this planet is not between animals and plants; it is between prokaryotes (bacteria without membrane-bounded nuclei) and eukaryotes (all the others made of cells with membrane-bounded nuclei). The detailed story of this huge discontinuity is connected to the origins of species.

The story of cell history is the story of how cells began to speciate. To see how, let us return to examine in more detail a subject deeply related to the origins of species—the rise in bacterial descendants accompanying the rise of speciation. For organisms that reproduce by fertile matings, species are real units easily described. Chimps mate with chimps, corn flower ovaries are fertilized by corn pollen, and frog sperm flows over frog eggs and converts them into fertile eggs that form tadpoles. Only members of the same species copulate, fertilize, spawn, or whatever and produce new members of their own species. But before the first nucleated cells the Earth harbored only bacteria, and with bacteria the apparently simple rules of the mating game are distinct. First, whether one-celled or many-celled, all bacteria reproduce all the time without any sex act at all. Moreover, many—if not most—bacteria in nature are multicellular organisms. Second, when bacteria do engage in sex they are promiscuous. They can all mate in some way (pass and receive genes) with bacteria of totally different kinds. And all this coupling and exchange of genes involves genetic recombination but not reproduction.

Here is the point. To understand the origin of species we need to agree on what they are, but we are very inconsistent. Zoologists and botanists are often satisfied with the rule of thumb that members of the same species produce fertile offspring, whereas members of disparate species do not. But more often than not botanists, zoologists, and all other biologists assign organisms to species based on morphology: the overall appearance and behavior of bodies. Bacteriologists can use morphology only in the case of the larger, conspicuous, or distinctive bacteria like myxobacteria, spirochetes, and many cyanobacteria. Otherwise they employ a different rule of thumb: If two kinds of bacteria share 85 percent of their measurable traits in common, they are taken to be members of the same species. If they enjoy 84 percent or fewer of their traits in common they belong to different species. Practicing bacteriologists are aware of this definition's arbitrariness, since bacteria change traits so rapidly. By this rule of thumb, bacterial species change all the time. Placement in a refrigerator or a warm incubator can cause some bacteria to change "species" in a few days. Zoologists, botanists, mycologists, and even protistologists would never agree that a week in a refrigerator is enough to change one species into another.

Our new definition of species, that organisms with the same number and kind of integrated genomes in common belong to the same species, depends on the recognition that all nucleated organisms are composite. All are products of integrated symbionts. The idea is that if two organisms (individuals) are members of the same species, then they are composed of precisely the same set of symbionts. Since bacteria are not formed by integration of symbionts, they lack species. The advantage of our analysis is that all nucleated organisms are assignable to species whether or not they ever engage in sex. If two protoctists share the same set of bacterial or other genomes—that is, if they share the same ancestors, made of unique sets of genes—by definition they belong to the same species. The same is true for two fungi, two animals, or two plants.

"Eukaryosis" refers to the drama itself, the evolutionary origin of nucleated cells. The first eukaryotes were protoctists. Protists, the smaller members of the protoctist kingdom, evolved by acquisition and integration of specific bacterial ancestors; the earliest protoctists by hypothesis were anaerobic protists in oxygen-free environments. Before eukaryotes, that is, before any protists evolved, the world was anoxic and entirely bacterial. No species existed in the Archean eon. The earliest eukaryotes, the first protoctists, were the first species on Earth—not, mind you, the first organisms. The acquisition and integration of alien genomes for the first time not only led to the earliest eukaryotes, it led to the first species in the history of life.

As casual relationships between bacteria with different capabilities became irreversibly intimate, these bacterial complexes generated the protoctist cell. The smallest eukaryotes, the earliest protoctists, led to the microbial ancestors of all large forms of life.

Many of the smallest protoctists, protists, by definition, live quite happily without any oxygen. Indeed, to them, oxygen is an instant poison. Among these oxygen-avoiding beings are the archaeprotists, the "amitochondriate protists." This phylum of hundreds of species lives in habitats such as the intestines of mammals and insects or in sulfide-rich muds. We believe these amitochondriate protists are living representatives of the ancestors of nucleated organisms (See Figures 7.4, 7.5 and 9.1). They cannot metabolize oxygen nor do they reproduce sexually, but they are clearly speciated. The two swimming species found in dry-wood-eating termites, for example, *Metacoronympha kirbyii* and *Trichonympha ampla*, can be easily and consistently distinguished.

If we accept the idea that species originated in eukaryotes prior to the appearance of either oxygen respiration or meiosis, the kind of cell division involved in sexual fertilization, we can resolve Darwin's dilemma of the origin of species from the beginning of the phenomenon of speciation itself. Familiar stable species can be named, identified, and classified once we agree that species themselves, and the process of speciation, exist nowhere in the bacterial world.

Speciation itself is thus a result of evolution. Not only are species not ideal, everlasting Platonic forms, but they only begin to

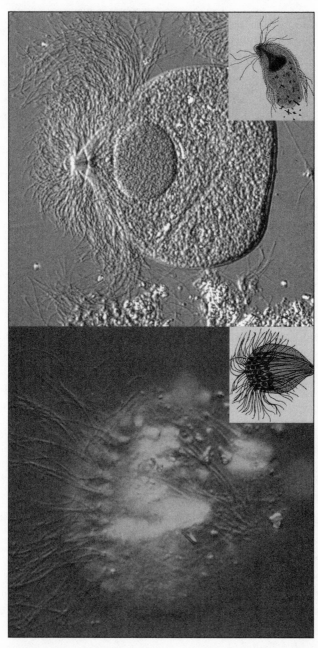

FIGURE 9.1 *Metacoronympha, Trichonympha,* Wood-Digesting Protists in Termites

exist in the last couple of billion years, after the origin of the nucle-ated cell.

Bacteria exist in a million guises. Some photosynthesize, many convert small organic compounds into myriad large ones. Some accumulate iron to produce magnets in their bodies. Some swim toward the light and away from acids. The world of bacteria possesses nearly every talent we associate with animal and plant life. New species appear not when organisms inherit "acquired characteristics" but when they acquire other genomes, entire other organisms with full sets of genes that determine the characteristics in question. Jean-Baptiste Lamarck was correct: Inheritance of acquired characteristics occurs, but only when the characteristics in question are determined by the genomes of acquirable organisms. Charles Darwin was also correct: New species appear when the newly acquired organisms are integrated, leave offspring, and are perpetuated by natural selection. The symbiogenetic origin of species is firmly grounded in the work of both Darwin and Lamarck.

DUBININA'S *THIODENDRON* AND DENNIS'S BLOOD

Though many will disagree with our conclusions, the previous chapters of this book are all based on well-established science. Here we tell a peculiar story in which the science is in progress and each detail not yet verified.

How did the nucleus, the distinctive and defining feature of the cells of all eukaryotic organisms, evolve? Probably not by mutation. The origin of the cell that is the common ancestor of all speciated life almost certainly involved "consortia bacteria," bacteria living with and inside other bacteria.

Once upon a time in the Proterozoic eon, approximately 2,000 million years ago, an organism called *Thiodendron latens* thrived at seaside locations all over the world. You can still find it here and there today. (See a summary of the geological "time rock divisions," the entire time scale, in Table 9.1.) This *Thiodendron*, which looks like white masses of stringy stuff and smells of sulfur, was studied by the Russian microbial ecologist S. A. Perfiliev in the mid-twentieth century. Indeed it was he who named it *Thio* (sulfur) *dendron* (finger) *latens* (slow or laid out) when he saw its filamentous cells stuffed with tiny globular yellow balls at intervals along their length. Some of the tangled filaments, very skinny but only four-tenths of a micrometer wide, were as long as one meter. Perfiliev had pioneered the placement of glass capillaries into muddy sediment not only to sample the layers of bacteria but also to measure gas exchange by the resident microbes. Back in the laboratory, when he grew great quantities of *Thiodendron,* he noted its capacity not only to produce the rotten-egg smell of hydrogen sulfide but also to oxidize the sulfide back to globular elemental sulfur globules. When oxygen penetrated the tangled culture, the yellow sulfur globules changed. Apparently the bacterium could oxidize the sulfur all the way to sulfate (chemically, the soluble sulfate ion, $SO_4^=$) which entered the water as it disappeared in solution. The filamentous bacteria cells, Perfiliev reported, frequently budded off small motile cells, flagellated swimmers that participated in the hydrogen sulfide production. In short, Perfiliev insisted that *Thiodendron*, like many other bacteria such as *Caulobacter* and *Chondrococcus*, enjoyed a complex life history in which filaments alternated with flagellated unicells. Unlike these other two genera, however, *Thiodendron* transformed sulfur from the sulfate ion ($SO_4^=$)to either the elemental form (S), the HSO_4^- thiosulfate form, or the reduced (H_2S) state. But then, if oxygen leaks in, this organism transforms whatever sulfur is around to the oxidized $SO_4^=$) state. The form of sulfur the organism produces depends on the availability of oxygen.

Perfiliev died before the powerful techniques of thin-section electron microscopy came into general use. He was succeeded by his student G. A. Dubinina, who, in 2002, at the end of her career as a microbiologist, still worked in Moscow University's institute of microbial ecology.

Table 9.1—Geological Time Scale*

	TIME-ROCK DIVISIONS			MAIN EVENTS	
	EONS	ERAS	APPEARANCES	CELL LEVEL EVOLUTION	TRANSITIONS
NOW	Phanerozoic	CENOZOIC	mammals	technoscience	**4**
		MESOZOIC	dinosaurs	Cretaceous extinctions	urbanization
		PALEOZOIC	shelled marine animals,	cell wall lignification	
			land plants	plant embryos and tissues	
					3 hard parts
	Proterozoic (Neo-Proterozoic)	UPPER	Precambrian-Cambrian transition	cell-to-cell junctions, animal tissues	
			abundant limestone		
			carbonate stromatolites, acritarchs	(sexual cysts) meiotic sex	
			fossils of first nucleated organisms	origins of nucleus from karyomastigonts chimeras, consortia	**2** eukaryosis
			banded iron formation, transition to oxygen-rich atmosphere	aerobic bacteria; oxygenic photosynthesis	
		LOWER	granitic rocks		
	Archean	UPPER	oceans basaltic rocks greenstone belts, tectonic activity	age of anaerobic bacteria	
			first stromatolites, microfossils and isotopic evidence of life	origin of life (bacterial cells)	
		LOWER	solid Earth-Moon system		**1** life
ANCIENT PAST	Hadean	NO EARTH ROCKS	meteorite bombardment, formation of Moon	meteorites and lunar rocks only of this age	

* simplified

Dubinina was curious to work out the developmental life history stages of Perfiliev's *Thiodendron latens* at high magnification. She collected *Thiodendron* from Perfiliev's original site north of St. Petersburg on the White Sea and also from the health resort of Staraya at the Lake Nizhnee mud baths near Moscow. For several years Dubinina and her colleagues attempted a modern description of *Thiodendron* based on electron microscopic analysis, physiological tests of sulfide production, and other laboratory work. In general Dubinina and various student and postdoctoral investigations confirmed Perfiliev's original description. But imagine their surprise when, in attempting to grow *Thiodendron* in a laboratory culture, they put the stringy sulfurous masses under fully anoxic conditions and the filaments, but not the swimming rods, transformed into long, skinny threads. They became spirochetes! The swimming rods, identified as the commonplace sulfate-reducing bacteria *Desulfobacter*, continued to swim around, hold their shapes and produce H_2S. The long filaments, by contrast, broke up into much shorter helical threads indistinguishable from standard spirochetes. This change required the complete removal of oxygen gas from the environs. If then exposed to tiny amounts of oxygen, the former filaments continued to oxidize it to elemental sulfur, which they deposited inside their cells. If the oxygen quantity rose, the spirochetes straightened out. Their cells remained very thin but became longer and longer. Unable to divide, the cells simply lengthened. Only after surviving for a long time in the stretched-out state did they finally die, presumably of oxygen exposure.

Dubinina's group went on to discover at least four more large clumps of seaside *Thiodendron*, each composed of very similar sulfate-reducing swimmer bacteria and slightly differing spirochetes. In the Pacific Ocean on Yankich Island at Kraternaya Bay and at Papua New Guinea, for example, *Thiodendron* was taken from bluish-white mats up to twenty millimeters thick. All, when placed in proper media without any oxygen, yield populations of standard free-living spirochetes. The authors invalidated the name *Thiodendron* when they made it clear that this was no single bacterium with a complex life history. Two very different types of bacteria had been isolated together: a swimming short rod that produced hydrogen sulfide and a filament-forming spirochete that thrived in the presence of the hydrogen sulfide it could not itself produce. Stable associations of entirely different bacteria are known as "bacterial consortia." The

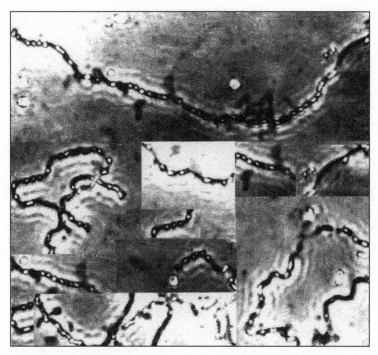

FIGURE 9.2 *Thiodendron latens* in Culture at Low-Oxygen Concentration

species name *Thiodendron,* Dubinina and her colleagues pointed out, had been given to a consortium of two. They went on to describe each member of the consortium. The first, a *Desulfobacter,* is the short, walled, flagellated sulfate-reducing hydrogen sulfide maker. It degrades sugar as it respires sulfate to sulfide. The other, the sulfide oxidizer *Spirochaeta,* converts sulfide to sulfur globules, which it stores in its body. The spirochete lives off the energy and food that ultimately comes from the long cellulose molecules of degrading algal cell walls, which are broken down by cellulase enzymes into a sugar called cellobiose. Many snails, fungi, and seaside bacteria make cellulase enzymes. The filament-forming spirochete thrives, like many, on cellobiose; it provides sugar breakdown products to the sulfate-reducer partner.

Dubinina's work was first published in English in 1993, in the Russian professional journal *Mikrobiologiya*. She and her colleagues pointed out that spirochetes can be clearly important geochemically to the transformation of sulfur in marine littoral zones. Only after we received a pack of photographs from Professor Dubinina did we realize all at once that we have seen *Thiodendron* many times. But we had always dismissed it as an ordinary seaside sulfur-oxidizing bacterium like *Thiothrix, Vitrioscilla,* or *Beggiatoa,* all of which tend to grow in white clumps (Figure 9.2).

No important evolutionary antecedent is ever extinguished without a trace. *Thiodendron,* we think, is a living legacy of the first step in the symbiogenetic origin of nucleated cells. What began as a consortium, extremely similar but not identical to *Thiodendron,* became a chimera, a new individual. Very different bacteria—one type that oxidized sulfide to sulfur, and another that reduced it back to sulfide—became fused. The two became one by symbiogenetic merger, and this chimera was the ancestor of us all.

For many years Dennis Searcy, a colleague at the University of Massachusetts, has argued that the "ground cytoplasm" of all nucleated cells evolved from a free-living, wall-less microbe comparable to one he grows in his lab called *Thermoplasma acidophilum. Thermoplasma acidophilum* is only slowly motile. Unlike *Desulfobacter* it is not flagellated, nor is it ever surrounded by a cell wall. Because the cytoplasm is not bounded by a rigid wall, *Thermoplasma* easily changes its shape. Dennis and his students have collected evidence that his *Thermoplasma* cells make "motility" proteins. These proteins underlie the phenomenon of movement. They permit the cells to slowly spread over the globules of sulfur Dennis has given them as their laboratory environment. The major difference between *Thermoplasma* and *Desulfobacter,* the partner of Dubinina's consortium, is that Dennis's organism lives in hot fresh water. It freezes at ordinary room temperatures, does not survive saltwater, and grows at acidities (pH 2.5 or so) far below those of the waters where *Thiodendron* was found (pH 8.0). *Thermoplasma acidophilum*, however, does produce copious amounts of hydrogen sulfide. Indeed it likes to live perched on bits of elemental sulfur globules out of the way of oxygen. It sticks to and even wraps around the surface of these globules. Unlike *Desulfobacter*, which is classified as a eubacterium, *Thermoplasma acidophilum* belongs to the archaebacterial subkingdom of prokaryotes. *T. acidophilum*, unlike nearly all other bacteria, produces histone-like proteins that surround its DNA. Dennis suspects these proteins protect its DNA from exactly the strong acid and high heat that would quickly degrade its genome (by "acid hydrolysis"). *Thiodendron*, on the other hand, does not need such extreme measures, since the sulfur-loving consortium lives bathed in neutral, cool ocean waters.

So how do these sulfurous bacteria connect to the story of the earliest eukaryotes?

We (Dennis Searcy, your authors, and perhaps a few converts) suggest that the earliest consortium that became the chimera that became the first eukaryote was a bacterial partnership similar to *Thiodendron*. We posit that a *Thermoplasma*-like archaebacterium that produced hydrogen sulfide (rather than *Desulfobacter*) and a gram-negative eubacterium, an organism very much like today's *Spirochaeta* that craved the hydrogen sulfide to protect itself from oxygen, formed an alliance. No doubt much juggling and mutual destruction followed before the partnership became fully permanent. Natural selection for speed swimming and organic food procurement forced these two types of bacteria into a tight association. Together they degraded sugars and other carbohydrates under sulfur-rich anoxic conditions. Probably the partnership thrived for millennia in fresh-to-brackish, warm-to-hot acidic waters, with the *Thermoplasma* analogue producing hydrogen sulfide that the other, the *Spirochaeta*, oxidized to elemental sulfur.

A student of Searcy's, Dean Soulia, set out to produce a *Thio-dendron*-like spirochete consortium in the laboratory using Searcy's *Thermoplasma acidophilum* with microbial mat spirochetes from our field collections. In the spring of 2002 this work is new. We know nothing yet about its success. The plan is to collaborate with Antonio Lazcano and Arturo Becerra of the Universidad Autónoma de Mexico (UNAM) as well as with Professor Searcy. Soulia so far has used Louis Pasteur's methods for growth of anaerobic microorganisms in the laboratory. He mimics the rules described by Charles Darwin, natural selection, by putting the organisms in oxygen-depleted hot acidic water that permits only the "fit" to grow and leave offspring. The goal is to obtain a stable consortium of *Spirochaeta* and *Thermoplasma* in a laboratory environment.

So convinced is Dennis Searcy that all eukaryotes carry around the legacy of their ancestors that he put his idea to the laboratory test. He sought cells that he could examine for the residual ability to produce hydrogen sulfide when given elemental sulfur. He chose his own blood cells because the red blood cells of mammals lack both nuclei and mitochondria. They resemble thermoplasmas in another feature: like all animal cells, they never have cell walls. Thus, Searcy figured, he wouldn't be confused by other cell organelles; he could just test the eukaryotic cytoplasm itself for an ability to metabolize sulfur, which no one has any reason to expect it to have. Not only did Searcy discover that his own cytoplasm generates copious quantities of hydrogen sulfide when supplied with elemental sulfur, he extended this observation to all four eukaryotic kingdoms. He gave elemental sulfur to yeast and molds and showed that fungi too make hydrogen sulfide (H_2S). The sulfide had to be removed as soon as it was made, as it tended to accumulate as poisonous waste. But once this precaution was taken, he easily demonstrated that all nucleated cells, whether plant, animal, fungal or protoctist, make hydrogen sulfide. Of course he could not remove mitochondria and nuclei from these other cells and test the cytoplasm by itself, as he had done with his own blood, but the work is convincing even so. Searcy maintains, and we agree, that our ultimate ancestral cells began as anaerobes in sulfur-rich environments. They still "remember" their origins because the ancient environment is built into the metabolism. The original consortium may just have been a *Thermoplasma* with a *Spirochaeta,* bound by sulfur oxidation and reduction, that permanently stuck together and then more tightly integrated over time. The eukaryotic cell formed by symbiotic alliance, the co-acquisition of very different bacterial genomes.

The next step in this chain, we think, was Radhey Gupta's chimera.

GUPTA'S CHIMERA

Professor Radhey H. Gupta doesn't drive anymore. His forearms have been so overused at the computer that he no longer trusts himself behind the steering wheel. Trained at the prestigious Tata Institute of Science on beautiful hills outside Bombay, Gupta came to Canada in the 1980s as an assistant professor of biochemistry at the McMaster University Medical School in Hamilton, Ontario. His interest is in deciphering relationships between microbial life forms, and his primary tool is the comparison of the long chain molecules all cells require for life: proteins and nucleic acids. Gupta focuses primarily on the long series of amino acid residues in proteins, primarily in bacteria. After detailed comparison of protein sequences in some 300 different bacterial types with similar proteins from anoxic protists, Gupta began to ferret out the history of the protist cell itself. Like us, he is especially interested in nucleated cells that lack mitochondria. We believe that some of these protists descended from ancestors that never had them.

An avid writer, Gupta has published many papers containing results and opinions that contradict the microbiological mainstream. The current mainstream scientist, for example, insists that all eukaryotes evolved directly from archaebacteria. No symbiosis or symbiogenesis was involved. Carl Woese, a microbiology professor at the University of Illinois, Urbana, and his colleagues have even renamed archaebacterial cells "Archaea," in denial of their bacterial nature and have elevated the group "Archaea" to parallel status with other bacteria and all eukaryotes, insisting on three fundamental groups of life. Gupta disagrees. All eukaryotes, he argues, even those without mitochondria at any stage in their life history, have a dual ancestry: two different prokaryotic ancestors. All nucleated cells contain some protein sequences that resemble those of archaebacteria (like *Thermoplasma*) and other protein sequences that resemble those of eubacteria (like spirochetes). Gupta insists that there is no reason to suppose that the eubacterial ancestor of the nucleated cell was more similar to a spirochete than to other eubacteria. (Although he disagrees with us about *which* eubacterial group was ancestral to all eukaryotes, he does not yet have a valid alternative to the *Spirochaeta* idea.) Still, use of the term "chimera" is Gupta's and the claim that all eukaryotes are chimeras is also his. Chimeras were ancient Greek mythical beasts with heads of lions and bodies of goats and parts of still other animals. They were composites whose body parts came from recognizably different origins. Gupta feels that description fits the cells whose protein sequences he studies. All eukaryotes, including those that lack both mitochondria and chloroplasts and live without oxygen or light, he asserts, are chimeric. The cells we think are today's closest relatives of the earliest eukaryotes are composite; they simultaneously have eubacterial and archaebacterial parentage.

Although as a biochemist Gupta does not use the terms "symbiosis" and "symbiotic partner," this is precisely what he is describing on the molecular level. His chimera was formed when two symbionts—one eubacterial, one archaebacterial—became fused in permanent liaison. We all agree that one symbiont was a eubacterium (whether spirochete like the partner in *Thiodendron* or some other) and the other symbiont was an archaebacterium (whether a sulfide-producer like *Thermoplasma* or some other). We—but not necessarily Gupta—believe the chimera had to be a good swimmer. We think the spirochete is a likely ancestor because it provided propulsion. The spirochetes that brought motility to the insides of nucleated cells propelled their thermoplasma partners to enter fresh, acidic, sulfur-rich waters. The thermoplasma, in turn, by production of highly reduced hydrogen sulfide H_2S, which "scrubs" oxygen gas, protected the spirochetes from oxygen, which at this point in Earth history (c. 2,500 million years ago) was becoming more and more significant as an environmental toxin.

As the fast-swimming spirochete became tethered to its slower sulfur-bound *Thermoplasma,* the composite cell, the chimera, evolved. In that chimera a new organelle system, called the karyomastigont, evolved in response to the imperative of symbiont integration. In the history of life, attachments that hold distinctive partner cells in permanent alliances have evolved many times. But even if we accept the bacterial consortium idea (*Spirochaeta* and *Thermoplasma*), how do we envision the transition from bacterial chimera to bona fide protist with its well-developed nucleus? The key is this cell structure, known from the beginning of the twentieth century. The karyomastigont, the nucleus, its proteinaceous connector to the cilium or "cell whip" (undulipodium), and the undulipodium itself comprise an organelle system found in many cells and usually ignored. Its evolutionary importance was only recently recognized by a colleague of ours, Michael Dolan. A former doctoral student at the University of Massachusetts, Dolan envisioned clearly why and how the earliest eukaryotic cells contained the karyomastigont with its "nuclear connector." This part of the karyomastigont is a protein structure that ties the organelle of motility, the former eubacterial spirochete, to the rest of the cell. The rest of the cell, in the opinion of Radhey Gupta and Dennis Searcy, was a former free-living archaebacterium. Searcy, who notes the way DNA is bound to protein and

how his cells move and attach to sulfur, would claim on these and
other grounds that the archaebacterial ancestor was like today's *Thermoplasma*. We agree.

The way we figure it, the membrane-bounded nucleus itself
evolved from a fusion of the chimera's DNA. The merger of the
once-independent *Thermoplasma* and *Spirochaeta* DNAs was like
any bacterial mating—the well-known passage of DNA from one
prokaryote to another. The archaebacterium *Thermoplasma* and the
eubacterium *Spirochaeta*, in typical bacterial-recombination fashion,
merged their DNA to make one single genome. The cells inside of
cells proliferated membrane. The new nucleus, which defined the
new chimera, was surrounded by membrane and was, from the beginning, tied to the organelle of motility, the former spirochete. The
spirochete evolved to become the undulipodium—a structure with
many aliases: cilium, eukaryotic flagellum, sperm tail. This moving
cell structure has such a well-known, characteristic pattern of tiny
tubules [9(2)+2] that all agree on its common ancestry. The organelle system (nucleus, nuclear connector, undulipodium composed of kinetosomes, and their axonemal shafts) minimally was
called the "karyo mastigont" (*karyo* = seed; *mastigont* = whip) by
those who first observed it. The Polish microscopist Robert Janicki
described it in tiny motile protists in 1933. Not until Harold Kirby
(1900–1952), a University of California zoologist and chairman of
the Zoology Department at Berkeley when he died, worked on the
karyomastigont did anyone really understand what it does.

We now think we understand how the nucleus evolved, mainly
because of Kirby's arcane studies. The nucleus originated as part of
the karyomastigont but then was released on its own. The reason we
think we understand nuclear origins is that certain live organisms
today behave just the way we think many of their ancestors did. We
must reconstruct evolutionary history from living clues that we take
to be representative. Evolutionary novelty of the nucleated cell is
best comprehended as specific historical products of partnerships

and symbioses, bacterial fusions of DNA whose products (proteins, RNA's, lipids) interact to generate emergent structures. Random mutations only refine and alter, but do not produce, species-level change. Protracted symbioses lead to symbiogenesis: the origin of new organelles, organellar systems, tissues, organs, organisms, and species. Symbiogenesis, the inheritance of acquired genomes, mostly those of bacteria and other microbes, is the greatest source of evolutionary innovation. Natural selection directs evolution through propagation and elimination of what it has already. Symbiogenesis, however, like the fusion of an archaebacterium with a eubacterium to form Gupta's chimera, is the big provider of the raw material that natural selection can then select.

LIBERATION OF THE NUCLEUS

The strange organellar system, the karyomastigont, common in protists and sperm, is composed of an undulipodium (at least one kinetosome-centriole [9(3)+0] and its [9(2)+(2)] shaft, the axoneme) connected by the rhizoplast or "nuclear connector" to the nucleus. A karyomastigont is shown simplified in Figure 9.3.

To understand how the cell nucleus evolved we must understand the karyomastigont and two of its related structures: the "akaryomastigont," shown in Figure 9.3, and the "paradesmose" (to be explained below). We did not create these ungainly terms to repulse the reader or to impede potential cell biologists. We do not introduce them here to make things more difficult. Rather, these terms, invented to describe what dedicated microscopists saw, date from the early twentieth century. Investigators at that time wanted to understand how bodies made of cells can grow larger, how cells "multiply by division." We still need the terms to understand how these types of cells arose in the first place. Janicki, Harold Kirby, D. H. Wenrich and their colleagues came up with these labels to describe moving structures in live cells and in stained preparations.

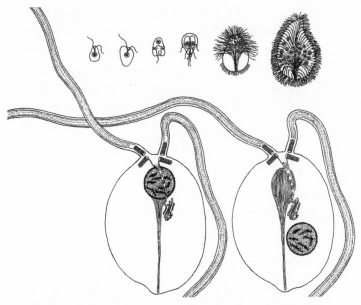

FIGURE 9.3   A Karyomastigont (Left) Compared with an
   Akaryomastigont (Right), (Selected Protists Useful in Reconstruction of Karyomastigont Evolution Depicted Above)

These investigators considered them prerequisite to understanding
individuality, reproduction, sexuality, gender, and other fundamental processes of living in our single-celled ancestors. Only now,
nearly a century later, can we make evolutionary sense out of the
karyomastigont, akaryomastigont, and paradesmose—the plethora
of microtubules, fibers, membranes, organelles, and entire organelle
systems of obscure protists.

   The bottom line is our idea that the karyomastigont (kineto-
some/centriole–nuclear connector–nucleus) is an "emergent" structure that first formed in the earliest eukaryote. Why? The organellar
system was selected for to prevent loss of the cell whip, the swimming organelle the chimera needs to quickly avoid oxygen and obtain food. There was a strong evolutionary pressure to attach the

genes of the original archaebacterium (like *Thermoplasma*) to those of its partner eubacterium (like *Spirochaeta*) so tightly that even death would not part them. The karyomastigont, in short, is the legacy of the original imperative of symbiotic partners to stick together and to be inherited together. The karyomastigont began with the membranous bag into which the two bacterial partners placed their genes—the nucleus. In some lineages the karyomastigont atrophied and left kinetosomes to proliferate at the cell's edge. In others the entire karyomastigont reproduced far more rapidly than its cytoplasm and generated new species of cells with eight or thirty or even 1,000 karyomastigonts. The last implies at least 4,000 kinetosome/centrioles, 1,000 nuclear connectors, and 1,000 nuclei. Each karyomastigont attaches with its nuclear connector to its single nucleus. In other lines of descent the karyomastigont reproduced faster than its nucleus could. New species evolved with hundreds of "akaryomastigonts" per cell. The "akaryomastigont" is the same organellar system as the karyomastigont but it has a big round space in the cell bounded by tubules and fibers where the nucleus should be.

In still other protist cell lines, such as the one that led to the *Snyderella* (Figure 9.4), the nucleus detached from its original karyomastigont and left akaryomastigonts to proliferate. Akaryomastigonts, in this lineage, each one with four undulipodia, presumably provide more swimming power to the cells that bear them. The result here was the evolution of a giant cell loaded with free nuclei and even more akaryomastigonts. Our evolutionary tale that the karyomastigont (tethered nucleus) preceded the liberated nuclei is supported by the observation of *Gyronympha*. This multinucleate genus, a relative of *Snyderella*, contains many akaryomastigonts and many free nuclei (unattached except at cell division). Almost as if it forgot to get rid of them, *Gyronympha* retains a few karyomastigonts (attached nuclei) as well. Life forms are exceedingly conservative. Once good body plans have evolved, they are retained.

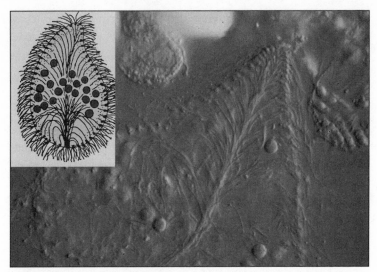

FIGURE 9.4 *Snyderella tabogae* (Many Unattached Nuclei and Even
More Akaryomastigonts)

Early investigators like Janicki, Kirby, and Lemuel Roscoe Cleve-
land studied what they called primitive animals, or protozoa, to discover how animal life evolved. They described the karyomastigont
with its attached nucleus and the changes as the "single-celled animals" grew and divided. First the "cell center," the kinetosome-centrioles at the base of the undulipodia, reproduced. Two became four,
four became eight. Most of the cells of interest were related to *Trichomonas vaginalis*, source of very annoying vaginal itch. They
tended to have four kinetosome centrioles at the base of four shafts.
In cell division, these four often became eight, with a thin fiber that
formed between the old and new kinetosomes. The fiber stretched
out in a line that the early biologists called a "paradesmose" (para =
alongside, desmose = connector). How the karyomastigont becomes
the paradesmose when the cell divides was beautifully shown by
Kirby. The nuclei, sometimes hundreds in a single cell, became

dumbbell-shaped. They pinched to double. Then the nucleus traveled up to the paradesmose and hitched a ride as the paradesmose lengthened. The elongating paradesmose finally moved the two offspring nuclei to opposite sides of the cell that becomes two cells. Without the stretched-out paradesmose, nuclei doubled but failed to move apart. Kirby saw the paradesmose as a kind of mitotic spindle. Like the mitotic spindle of animal cells, the paradesmose doubles its kinetosome-centrioles; it grows, elongates, and dissolves. The paradesmose, like the spindle, acts to segregate its offspring nuclei during cell division.

The electron microscope confirms Kirby's insight. It shows that the paradesmose is composed of standard 24 Angstrom-diameter microtubules (twenty-four nanometers in diameter), just like the spindle. The structure is simply a skinny, early version of the mitotic spindle. The karyomastigont of the nondividing cell and the paradesmose that it forms during division we take to be the legacy of eubacterial-archaebacterial symbiotic merger. The paradesmose is the protein structure that ensures that the formerly independent partner bacterial genomes will be inherited together. This is what made the alliance of eubacterium and archaebacterium permanent.

PART FOUR

CONSORTIA

CHAPTER 10

SEAWORTHY ALLIANCES

*D*onald I. Williamson, of the University of Liverpool, has spent his professional life at the Port Erin Marine Laboratory on the Isle of Man. There on his isolated marine bench, he has formulated "a saltatational process in animal evolution which operates independently of the accumulation of mutations and then selection."

Stripped down to its basic idea, Williamson's claim is that the major evolutionary changes in backboneless animals (and nearly all animals are invertebrates) emanate from inheritance of acquired genomes. He argues that the genomes that determine the larval animal forms are different from those that determine the adult forms—nor are they from the microbial world. Rather they are from other animals. Unlike the gaudily colored foreign genomes—usually green—that betray the presence of photosynthetic bacteria in algae and in plants, the colorless watery animal-animal genome transfer championed by Williamson tends to be overlooked.

Williamson explains that the origin of larvae, the immature forms of insects, starfish, and so many other marine reef and mud dwellers, bespeaks bizarre and arcane symbioses so integrated that only obscure clues remain. "I dispute the widely held assumption that larvae and corresponding adults have always evolved together within the same lineage, and I present my alternative hypothesis of larval transfer," he wrote in 2001 in the *Zoological Journal of the Linnean Society*. "[My] hypothesis was born of the conviction that evolution entirely within separate lineages is inadequate to explain the distribution of larval types in the animal kingdom, and also that the methods of metamorphosis which link successive phases in development could not have evolved merely by natural selection of random mutations. I emphasize, however, that I do not propose larval transfer as a substitute for natural selection. Adults and larvae have evolved gradually by 'descent with modification,' but, superimposed on this process, entire genomes have been transferred by hybridization."

Williamson posits that sexual encounters, by either external (eggs shed and fertilized in water) or internal fertilization (penis-aided entry of sperm into females), sometimes occurred between individuals of very different classes and phyla. Occasionally they were spectacularly successful. Such successful matings between very distantly related animals occurred infrequently, some thirty to fifty times in 541 million years. This means a fertile, successful outcome happens roughly once in 10 million years. The occurrence during a development of two or more body forms (larva of one kind that develops into distinct kinds of adult) he takes as direct evidence for the presence of separate but integrated genomes (heterogenomes).

The data that led Williamson to his radical suggestion occupy much space in the literature of marine zoology. Extremely different adults (sea urchins, brittle stars) enjoy nearly identical larvae (called pluteus), whereas closely related adults (a second species of a sea urchin genus or a starfish) develop from entirely different larvae (bipinnaria). Williamson cited the great zoologist Libby Hyman of the University of Chicago and her forty-volume work on fifty phyla of animals, published in 1940. In describing members of the phylum Coelenterata, she wrote, "Often closely related hydroids produce very different free medusae and vice versa, so that the medusae may be classified into a different family from the hydroids which bore them."

Fertile sexual encounters between adult animals of very different parentage were even witnessed in at least one case: sea urchin fathers (*Echinua esculentus*, animals in the phylum Echinodermata) were seen by Williamson to fertilize chordate eggs. (The Chordata is *our* superphylum. Today familiar animals with skulls are placed in phylum Craniata within the superphylum Chordata. The Chordates include all the backboned "charismatic megafauna"—tigers, alligators, Galapagos tortoises, elephants, sharks, giant condors, and so on.) The chordate mother that had eggs fertilized by a sea urchin father was a sea squirt named *Ascidia mentula*. Not only did the fertilized eggs survive this bizarre coupling but they developed fully paternal larvae, the immature forms called plutei. Unlike normal sea urchins, however, most of the pluteus larvae from the anomalous chordate eggs failed to develop normal adult organs. Rather they retracted their larval arms and produced spheroids: The pluteus larvae developed into rounded animals each with an adhesive disk. Such developmental behavior is unknown in normal *Echinus* fertilized eggs. But adhesive disks are always produced by *Ascidia* sea squirts. The developed disks allow the normal mother, the adult *Ascidia*, to adhere to rock or other solid substratum after it settles down from its tadpolelike larval phase. Some of these weird cross-phylum hybrids survived up to ninety days after hatching. A small number of larvae developed rudimentary organs typical of the father sea urchin and did, after thirty-seven to fifty days, develop into normal-looking sea urchins as if their chordate inheritance had disappeared entirely. Four years later, the three surviving urchins produced eggs. Williamson fertilized these hybrids' eggs with wild sea urchin sperm. Normal pluteus larvae resulted. Even their ribosomal and mitochondrial DNA was typical of *Echinus*. Apparently only the paternal echinoderm genome had survived in the animals that developed into the adult sea urchin form. But in the spheroid-forming majority, the *Ascidia mentula* sea squirt genome held its own. This weird cross-phyla hybridization is analogous to ordinary Mendelian breeding experiments within a single species where the trait that appears in the hybrid reflects the dominant (and not the recessive) gene. Only in sea urchin *Ascidia* cross-phyla matings a complete haploid genome, crossed with a different haploid genome, reveals which entire genome is at first dominant: that of the sea urchin. But development ensues such that "genome dominance" is related to state in life history.

Here in the cross between sea urchin male and the ascidian female the sea urchin paternal complete haploid genome predominates and the female genome disappears. Although attempted, the reciprocal genomic cross, the sea urchin female and the ascidian male, has never been successful. A great opportunity to study illegitimate mating, that is, cross-phyla hybridization, awaits an enterprising invertebrate zoological geneticist—or shall we say genomicist?

The reality of the genomic cross is clear to Don Williamson. He has witnessed hybridization between animals who are not only not of the same species but are not of the same genus, family, order, class, or even phylum. The strong inference is that genetic fusions like these some 225 million years ago occurred between echinoderms such as certain starfish. Certain starfish even today lack any larval stage but develop as adults straight from eggs. Other starfish, in Williamson's reckoning, acquired new complete genomes by sexual merger. Several types of adult planktic swimmers were acquired and became larval forms of starfish that previously lacked any larval stage. Indeed, all echinoderms, according to Williamson, lacked larvae until just after the greatest mass extinction in the history of life: the Permo-Triassic that occurred 225 million years ago. As the

Mesozoic era began, the fertile echinoderm survivors of several different classes independently acquired and retained adults with which they had illegitimately mated. This tendency toward bizarre taste and promiscuity has not entirely disappeared, as Williamson has shown. Extant echinoderms might still be capable of larva acquisition from urochordates such as sea squirts with their nice tadpole larva. They sprout adhesive disks when it is time to settle into adult living.

Williamson has expanded his theory. In his 1992 book about incongruous larvae, he theorized that only certain species, in eight phyla, had acquired their larval stages by transfer. Now he claims that all species that produce larvae, even caterpillars and other terrestrial animals, acquired foreign genomes at some point in their history. Williamson cites highly complex developmental pathways from multiple larvae to adult, for example in decapod shrimp belonging to the family Sergestidae. These crustacea metamorphose four times during their development from egg to adult ten-legged shrimp. All four larval types are planktic, which makes it difficult to see how random mutations accumulated to generate them in a uniform environment. (Benthic protoctists and animals—those living on the ocean floor—tend to be far more varied than their planktic relatives both now and in the fossil record.) Many unperturbed sergestid shrimp die at metamorphosis, which reflects intense selection on the hybrids, but those who survive transform from egg to naupliomorph to plenocarid to mysidacean to mastigopoid to adult sergestid. This plethora of larval transformations in single decapod shrimp speaks to Williamson about the persistent presence of the past. Sergestid shrimp, he suggests, acquired, integrated, and put to work at least four intact genomes. To Williamson the inheritance of these acquired genomes, not random mutations, determines the evolutionary success of these shrimp today.

In seminars, Don Williamson likes to show a transparency of Ernst Haeckel's familiar nineteenth-century animal family tree. The tree represents only diversification—branches split to form more branches. No symbiotic fusions are depicted and certainly no crosses between animals assigned to different phyla. Of course, Williamson concedes, Haeckel was a highly creative, dedicated scientist. "But why," he asks, "when he designed his phylogenies based on evidence garnered before the 1880s, does Haeckel still have to be correct about his basic premises?" The question, like the projection slide of Haeckelian zoology, lingers in the darkness of the seminar room. "Take a good look at this phylogeny," he tells his small, intense audience of zoological faithful. "It is about to go extinct."

After studying larva in the sea, the laboratory, and the library, Williamson has collected many reasons to doubt that marine animals of vastly different types evolved by random mutation from extraordinary similar—if not identical—larvae. He rejects Haeckel's tree and the neodarwinian style of thought. He instead posits genome acquisition, integration, and inheritance among survivors as the rule in the evolution of animals with anomalous larvae. The sources of larvae, Williamson concludes, are several. They include rotifers, which gave rise to the trochophore larvae of annelids and mollusks. Adult onychophorans, called "velvet worms," look suspiciously like caterpillars. Perhaps, Williamson suggests, they provided the larval genomes for lepidopterans (moths and butterflies), certain neuroptera (scorpion-flies), as well as most hymenoptera (bees, wasps) that are assigned to the suborder Symphyta (See Figure 10.1). The other suborders of Hymenoptera, which have entirely different larvae, presumably did not make the same mate choices. Members of several different insect orders acquired onychophora genomes by hybridization.

Tadpoles expand Williamson's armamentarium. Larvaceae (appendicularians), in our chordate superphylum and are classified with the ascidians as Urochordate (tail chordates), do not metamorphose at all. They remain as tadpoles all their lives, including their own reproductive lives. Tadpole larvae, claims Williamson, were ac-

Peripatus (adult)

Panorpa (scorpion-fly caterpillar larva)

Nematus (wood-wasp caterpillar larva)

Pieris (butterfly caterpillar larva)

FIGURE 10.1   Williamson's "Transferred Larvae"

quired when adult larvaceans successfully donated their genomes, probably by hybridization, to the chordate sea squirts (they have peculiar fishlike larvae). In general, in Williamson's view, adult metazoa that developed directly from eggs preceded any larval forms in animal evolution. This view is testable: If correct, the fossil record should routinely reveal that adults precede larval life histories. Furthermore the same kinds of larvae have intercalated themselves into different lineages at different times in the Phanerozoic eon. Some larvae should have been acquired by foreign lineages even quite recently.

Williamson strongly urges that researchers verify the larval transfer theory by molecular biological study of these genome acquisitions. Our opinion is that his "radical" ideas deserve careful and intense scrutiny. If he is correct, then the branches of animal evolutionary trees do not just branch but fuse—they anastomose like those of the frankly symbiogenetic lichens and photosynthetic animals. Animal evolution resembles the evolution of machines, where typewriters and televisionlike screens integrate to form laptops, and internal combustion engines and carriages merge to form automobiles. The principle stays the same: Well-honed parts integrate into startling new wholes—like the six shockingly different stages displayed from the egg to the adult of the sergestid decapod shrimp.

NEMATOCYST THIEVES

We will examine only two more examples of inheritance of acquired genomes in marine animals: in both cases the details of the merger help us understand the potential for saltatory evolution, rampant speciation, and appearance of genuinely novel taxa. We look at hydramedusoids in the phylum Cnidaria and, in the mollusks, at cephalopods that grow directly from eggs and glow in the dark: *Euprymna scolopes*.

To the phylum of Cnidaria belong the freshwater hydras, the Portuguese man-of-war, the siphonophores of the deep sea vents, and the flower-like anemones of tide pools. All members of this ancient, far-flung division of marine animals have stinger cells, activated by triggers that make the phylum infamous. These stinger cells, called nematocysts, are complicated organellar systems packed inside specialized cells called cnidocysts. Comb jellies (ctenophores) for many years were grouped with hydromedusans into the single phylum called Cnidaria, primarily on the basis of the similarity of the nematocysts. Since comb jellies, with their long trailing extensions filled with densely packed cilia, have little else besides stingers in common with the alternating-generation hydromedusans, the two phyla have again been separated: Coelenterata (hydramedusoids) and Ctenophora (comb jellies).

Elie Metchinikoff, the Russian discoverer of bacteria-eating macrophage cells in the blood of crabs, spent his summers in southern Italy at the straits of Medina, which pass between Regio Calabria and Sicily. During the late nineteenth century a fine marine station built by the czarist navy was maintained there. Like many pre-Soviet biologists, Metchinikoff could pursue his research from a European holiday site. Russians and other northerners, before the Great War, contributed mightily to marine biological explorations from such choice summer locations.

Metchinikoff's brilliant contribution to science, derived primarily from his study of crab phagocytic cells, included not only the cellular theory of immunity. He also insisted that a surprising lack of correlation existed in the classification of hydrozoans. These coelenterates resemble *Hydra* but their life history also includes the "upside-down bowl" medusa stage. The hydroid stage of some correlates perfectly with the medusa stage of the same animal, whereas in others these two stages don't match at all—the medusa looks like it belongs to a different species altogether. This incongruence between adult and larval classification, Williamson suggests, is just another case of acquisition of one set of genomes, the hydroids, by another set, the medusoids. The alternation of generations, hydroid with medusoid, is entirely diploid: In all the life histories both the upright tentacled hydroid and its alternate, the bowl-shaped, downward-facing medusoid, are diploids. All their body cells, in other words, bear two sets of chromosomes. Williamson claims that his idea that these life histories have not always been together helps explain why some hydrozoans entirely lack a medusoid stage while other "hydromedusae" have no medusa stage at all. It also explains why, in the same single individual coelenterate (one with both life history stages in its development), one sees the medusoids move separately from the hydroids. This separate behavior implies separation of the nerve connections, as though two different complete nervous systems were simultaneously present in the same individual. "The two body forms originated as the genomes of two once-independent animals," wrote Williamson. Whereas budding in a single body form of hydroid or medusoid implies the usual single diploid genome, the occurrence during development of two distinct body forms, especially with uncorrelated muscle movements, is taken by Williamson as an argument for fusion of two formerly separate genomes.

Whereas this thesis about coelenterate double origins is still debatable, the probability that the stinging cells, the nematocysts of coelenterates, were free-living organisms whose genomes were independently acquired by comb jellies and hydramedusoids seems to us likely. The candidate for the original free-living stinger is a microsporidian, well described in 1995 by Stan Shostak and Victor Kolluri.

Speciation, and even the origin of more inclusive taxa, in these "jelly animals" seems best understood as yet another effect of the inheritance of acquired genomes.

The surfaces of hydroids that lack medusoid stages, such as tubularians, are typically studded with stingers. The term nematocyst refers to the stinger cell wherever it resides. The oval nematocysts have sharp barbs and a poisoned tube folded inside. Each cell is triggered. When the trigger is mechanically stimulated the barbs and poisoned tubes are released to the distress, indeed often the death, of the prey or potential predator. Most sea creatures leave the well-protected hydroids alone.

Naked slugs, shell-less mollusks called aeolid nudibranchs, feed on stinger-studded hydroids without peril. In fact they eat little else. Often their coloring resembles that of their food. Aeolidae members of this slug family flaunt their stolen goods. Indeed the slugs flaunt the stolen genomes. The shell-less snails ingest hydroid parts and digest away body parts they don't need. The slugs devour the entire tubularian body except the stingers. The slugs then detrigger the stingers so they never sting—it is like swallowing a loaded pistol. The complete nematocyst cells travel down the slug's digestive tract like any other food, but unlike food the stingers somehow resist digestion. They end up in outpockets of the slug gut. The stingers are then stored in translucent distal portions of the slug intestine in waving, conspicuous protrusions. The highly modified "nematocyst holders" are called cerata. One single finger-shaped projection is called a ceras. Of course the detriggered nematocysts never fire inside their living molluscan traps. No one knows for sure how the slugs store the foreign weapons detriggered in their own bodies without accidental firing, or how the slugs choose among the nematocyst types for just the ones that suit them. But what everyone does know is that the bright colors of the slugs, with their waving cerata, are fair warning to those who attempt to ingest them. The orange cerata of *Hermisssenda crassicornis* (Figure 10.2), the beige protrusions of *Cratena*, or the brown spots of *Eubranchia* are instantly recognizable. All the members of the gastropod suborder Aeolidae play the nematocyst game. They all feed exclusively on nematocyst-studded hydroids that no one else dare eat. No possible slow accumulation of mutations could account for this adaptation.

Rather the evolutionary engine was acquisition and cyclical reten-
tion of a foreign genome, the genome of the hydroid weapons.

The molluscan subclass Opisthobranchia, in which the suborder
Aeolidae is placed, seems particularly suited to stealing the genomes
of others. The Ascoglossa, another order of gastropod mollusks, tend
to be green. They steal genomes of photosynthesizers, not nematocysts. A survey of eighty-six species showed 82 percent to be genome
thieves colored green. Native to salt marshes from southern Florida
to Nova Scotia, *Elysia chlorotica* retains photosynthetic plastids from
Vaucheria litorea. If starved for external food, this slug can live pho-
toautrophically for months. *Elysia timida* feeds only on young
Caulerpa, a huge, green, single-celled alga, and retains its chloroplasts. According to the scientist who studied it, the survival of this
Elysia "is dependent upon adapting to the life cycle of the chloro-
phycean alga *Acetabularia acetabulum.*" *Elysia tuca* tolerates calcified
algae and collects its chloroplasts from the noncalcified portions of
Halimeda incrassata. Here again we see the correlation between recognized species and the kinds of genomes they retain.

GLOW-IN-THE-DARK SQUID

The acquisition of a foreign genome, that of the gamma proteobacterium *Vibrio fischeri*, changes a small squid into a nocturnal glow-
in-the-dark escape artist (Figure 10.3). Here the genetic relationship
is powerful: No squid of this species lacks a full-blown complement
of bioluminescent (cold light) bacteria. Yet we would be stretching
the facts to call this the "inheritance of acquired bacteria" because
the association is cyclical. Squids hatch from eggs without bacteria.
They flush out 90 percent of their bacterial inhabitants every day.
What is inherited from parent to offspring squid are animal genes
that control bizarre tissue development: a huge two-lobed spotlight
in the squid's belly, a special epithelial core, a modified translucent
muscle, and a dark reflector that directs the cold light. The squid's

FIGURE 10.2 *Hermissenda* Nudibranch, a Nematocyst Thief

body has undergone massive transformation, over evolutionary time, to become a receptacle for its glowing bacterial inhabitants. But if the correct strain of live potential bacterial tenants is absent from the environment, the entire animal light organ fails to form.

Margaret McFall-Ngai and Ned Ruby keep a colony of ten to twelve mated pairs of squid in their laboratories at Kewalo Marine Station at the University of Hawaii. These captives lay about 30,000 eggs a year. If trouble occurs in the squid tanks all McFall-Ngai and Ruby need to do is to walk out the door of the Pacific Biomedical Research Center to the beach and take samples from the sea water.

Unlike many of Williamson's mollusks, *Euprymna scolops* has no larval stages. Embryos develop for about twenty days in the ocean water and, with no parental care, juvenile squid hatch just after dusk. All new hatchlings bear only immature light organs.

The baby squid's light organ is covered with cilia. A ring of beating cilia forces water toward a series of pores, three on each side of the light organ. Each pore leads to a duct and thence to a crypt where the bacteria settle in. The squid light organ, with its welcoming ring of cilia, is ready for its bacterial residents as soon as the squid hatches. Within twelve hours the squid respiration and ciliary movement has brought in enough bacteria that they line the intercellular

FIGURE 10.3 *Euprymna scoleps,* the Tiny Luminous Squid

spaces. The bacteria population begins to grow by division until it fills the organ. Only the correct *Vibrio fischeri* populate the *Euprymna scoleps* light organ. Somehow, the squid is more adept than the most accomplished microbiologist at identifying and cultivating its own proper bacteria from the confusing mixture in seawater. Some twelve hours after moving in, when the animal is about a day old, the bacteria send a massive and irreversible death signal to the squid tissue that hosts them. The epithelial cells of the ciliated surface die on command. Some four days after hatching, all traces of the "welcome" organ have disappeared. The signal is subtle and indirect. The bacteria must enter the light organ crypt by swimming. They must induce a fourfold increase in the volume of the squid surface epithelial cells. The number of microvilli, little projections, on the top surfaces of the crypt cells must dramatically increase. The game is over if the bacterial symbionts are removed by heat or antibiotics. The light organ recedes to its status quo ante to resemble the organ found in the original hatchling. Bacterial genes that code for "adhesin" proteins on the bacterial cell wall surfaces seem to promote the light organ's maintenance.

Why are the genetic systems of both the squid and its resident bacteria so keenly tuned to the development of this huge light organ? Apparently "counter-illumination"—a kind of camouflage—is the reason. Predators from below or potential prey see night sky instead of the tasty little squid belly when they gaze at the mature illuminated squid. The bacterially generated ventral light is the same intensity, color, and angular distribution as moonlight. And as starlight. Whereas during the day the bacterial-bloated squid sleeps in the sand, during its night-time feeding sessions the squid goes abroad counter-illuminated by its bacterial symbionts. Experiments show that squids without this counter-illumination are immediately vulnerable to predators. Although such squid live well if protected in the laboratory, *E. scoleps* does not exist in the wild without its symbionts.

FORTEY'S OLENIDS

The inference of speciation and the evolution of higher taxa by symbiogenesis is often fraught with difficulty even with live organisms available for analysis. We must be even more tentative in claiming to find symbiogenesis in the Cambrian fossil record. But the English paleontologist Richard Fortey has done just that. Trilobites of the family Olenidae (the olenids) left abundant carapaces (body coverings) in black shales between the late Cambrian and the end of the Ordovician (Figure 10.4). Trilobites, a class in the phylum Arthropoda, were jointed-legged but, unlike their modern insect and crustacean relatives, not chitin-covered. Their carapaces were constructed mainly of calcium carbonate. These calcium carbonate shielded animals filled the Paleozoic seas with charming diversity. Some were as tiny as flies, others as large as huge lobsters. Some hardly moved, others were swift, ferocious killers. By the Permian period, 260 million years ago, they were all extinct. They had lasted a glorious 300 million years—one hundred times longer than the species of man, at least so far.

Olenid trilobites thrived on sea bottoms worldwide for some 60 million years. Some sixty-five different genera, a total of over one hundred species, have been named. The most intensively investigated of their fossils come from sediments in Norway and Sweden. Olenid fossil remains tell us these animals must have covered the sea bottom in droves from what today is northern France to northern Norway.

Virtually all trilobites bear a strong ventral plate called the hypostome, which functioned in the manipulation of food. The earliest, simplest, and supposedly the most primitive of the olenids, *Olenus* itself had a normal hypostome. But the ventral plate is atrophied in the majority of fossil species. In one subfamily, the Pelturinae, the mouth plate has entirely degenerated. Articulated, undisturbed molts abound. These and other indicators suggest the

FIGURE 10.4   Richard Fortey's Trilobites: *Olenus* in Their "Nobody-at-Home" Beds

olenids thrived in a habitat of deep quiet water. Furthermore the molts indicated the bodies themselves were flattened, with feeble axial muscles. Apparently they lived in places that lacked much bottom current. The olenid carapaces are found pyritized—calcium carbonates were embedded with iron sulfide, that is, pyrite, also known as fool's gold. Pyrite forms immediately if sulfide (H_2S gas) is brought into contact with iron-rich water. Fortey figures the olenid trilobites must have been bathed in sulfide-rich, oxygen-poor, sluggish water.

Unlike most of their trilobite brethren, olenids have exceedingly thin cuticles. They bear comblike lamellate structures called "exites." Some paleontologists interpret these filigreed structures as "gill branches"—coverings of the extensions (gills) that "breathe," or exchange oxygen gas dissolved in waterlike fish gill coverings. These gill branches no doubt greatly increased surface contact with sulfide-rich sea water. Fortey, who works from early morning until late at night on these fossils at the British Museum (Natural History), Kensington, London, suggests that all the olenid trilobites harbored bacterial symbionts. As with modern deep sea vent tubeworms (vestiminiferans), several genera of annelids (segmented worms), and bivalve mollusks that have extremely well developed symbiotic associations, with trilobites we can take morphological changes as a guide to symbiotrophy. Living animals that cultivate such bacterial symbionts in their tissues show distinctive signs of the association. Like Fortey's fossils, many have modified mouth and other digestive parts. Some bear hypertrophied trophosomes (organs that store bacteria) or gill lamellae. Furthermore the bacteria-associated animals tend to reproduce in large numbers.

Some olenids have spectacularly large brood pouches. Fortey suggests that larval (baby) trilobites were protected inside until they cultivated enough carbon-dioxide-fixing, sulfide-oxidizing symbiotic bacteria to become self-supporting. In the Great Quarry at Andrarum, Sweden, the "Alum shales" of the Upper Cambrian (505 million years ago)—the so-called *Olenus* zone—has been investigated in detail by Euan N. K. Clarkson of the University of Edinburgh. A clear correlation between the presence of pyrite, iron sulfide, and two olenid species, *O. truncatus* and *O. wahlenbergi,* can be traced for one-and-a-half meters of a shale section. Where pyrite disappears, so do the olenids. A fossil ostracod appears to take their place. *Cyclotron,* a kind of marine crustacean, by inference avoided high sulfide waters and lived where oxygen made life tolerable. It dominates where it could apparently breathe oxygen. The olenids lived out of the way of any oxygen leaks. These ancient trilobites, Fortey believes, built the carbon of their bodies through the autotrophic favors paid them by the carbon-dioxide fixing bacteria that inhabited their bodies. The olenids cultivated bacteria, they "body-farmed" just like Andrew Wier's *Staurojoenina* (see Chapter 7) and Margaret McFall-Ngai's Hawaiian luminous bobtail squid. The sulfide-oxidizing bacteria still fix carbon the way plants and algae do, but they do not grow in the presence of oxygen gas. In Fortey's olenid these bacteria may even have existed in pure culture inside the trilobite's body. Thus 500 million years before Louis Pasteur and without even the benefit of language, olenid marine arthropods seemed to have evolved the capacity to culture bacteria. The cozy arrangement saved the trilobites the bother of feeding. Sulfide-oxidizing bacteria served the olenid's own reproductive imperative as this great suborder generated its sixty-five different genera and perhaps one hundred species. Under sulfur-rich, oxygen depleted conditions the olenid's fellow trilobites, lacking the symbiotic bacteria, would have suffocated and starved to death.

CHAPTER 11

PLANT
PROCLIVITIES

POOR MAN'S UMBRELLA

Fertile "sexual" relations, with bizarre, theoretically barren couplings, occurred throughout the history of life between land organisms of different species—and continue today. A huge-leaved herbaceous plant, *Gunnera manicata*, thrives in the cloud forests of the Andes. In Ecuador, at about 20,000 feet, this "poor man's umbrella" maintains a partnership with *Nostoc*, a cyanobacterium (Figure 11.1). *Nostoc* helps confer nitrogen, always in short supply, on the rest of the plant community. It happily infects stem glands; its filaments glide into special tunnels of the leaf stalks. There, in collaboration with the plant, this cyanobacterium performs symbiotic nitrogen fixation inside its heterocysts. *Gunnera* is by no means the only sedentary land dweller thus fertilized by a cyanobacterium. Such odd couplings abound.

Today nearly 10,000 different cyanobacteria can be distinguished by a very few experts. Before any algae or plants evolved

FIGURE 11.1   *Gunnera manicata* and Its Symbiotic Cyanobacteria

probably many more types of cyanobacteria existed that now, alas, are extinct, like most organisms from the past. When one or a few kinds of cyanobacteria were spared digestion by plant ancestors and became the chloroplasts, they left most of their relatives outside. The internalized cyanobacteria either lost or never had the ability to "fix" nitrogen, to remove the chemically refractory N_2 gas from the air and incorporate it into the organic compounds of their bodies. To gain nitrogen, we animals eat beans or meat. We take our fixed nitrogen from our food. We suffer nitrogen deficiency if we fail to eat well. Even though we are bathed in an atmosphere that is almost 80 percent nitrogen, in no case can we compensate for dietary nitrogen deficiency by grabbing the needed element from the air. This talent, the incorporation of gaseous nitrogen into carbon-rich chemical compounds of the body, belongs solely to bacteria. The metabolism, the body-level know-how, is simply beyond us. It never evolved in any other kingdom of organisms, but several types of nonphotosynthetic bacteria know the trick. They all contain the "nitrogenase" set of genes that permit fixation of extremely refractory gaseous nitrogen into water-soluble amino acid compounds.

The 30 million species of fungi, animals, plants, and protoctists were all forced to acquire the nitrogen of their proteins in another fashion. Some overcame their persistent nitrogen deficiency by literally cuddling up with bacterial nitrogen fixers and acquiring genomes on either a temporary or permanent basis. The consequence of this genome-coveting behavior led to new taxa: new species, new genera, and in some cases new families and classes. We only describe a few of the evolutionary novelties here to drive home our point: Evolutionary innovation never depends solely on the accumulation of random mutations or even duplications of well-honed genes. In the cases mentioned briefly here, the taxa in question, new fungi and plants, evolved when the fungus or plant acquired and stably integrated part-time or full-time the genome of a nitrogen-fixing bacterium.

THE SECRET LIFE OF *GEOSIPHON*

Geosiphon pyriforme, a few millimeters high, was discovered in Neustadt, Germany, by a secondary school science teacher in the early 1800s. He took the "plant" from along the banks of a river in damp soil near the water and placed it next to the name "*Botrydium*" on herbarium paper. These tiny pear-shaped bladders in clusters of three to ten or so are extremely rare today. They have no fossil record and in fact have only been studied in Germany and central Europe. None have been seen in either North or South America, or the Pacific. A fine botanist, Fritz von Wettstein, relabeled and renamed the organism on the herbarium sheet when he discerned its true nature. Now, after many years of laboratory and field study, D. Mollenhauer has brought to light the secret life of *Geosiphon*. It is a double organism, rather like a lichen, with two different types of ancestors. One is a fungus that resembles *Endogone*, *Glomus*, and other well-known members of the fungal phylum Zygomycota. The other ancestor is the cyanobacterium *Nostoc* that fixes nitrogen. The fungus becomes green, enlarges to a "monstrous" (for a microbe) size of about forty-four millimeters and makes peculiar bladders unknown elsewhere in botanical literature.

Both the fungus and the *Nostoc* change radically in the association. They fix the two atmospheric gases of importance to their bodies: carbon dioxide (via photosynthesis) and nitrogen (via nitrogenase). These useful fixations occur only in the integrated "plantlike" being. Under the influence of the fungus the gliding propagules of the cyanobacterium, called "hormogonia," are trapped. They become bloated and stop their motion. After it becomes clear that the captured cyanobacterium cannot get away, a magical transformation occurs. The fungus cell wall breaks, its membranes open to the outside. A new wall and membrane form at the breach. The fungus stuffs the cyanobacterium through the opening, which then totally closes. In a remarkable metamorphosis the partnership matures into what looks like a plant. The bladders fix nitrogen and carbon actively for a number of weeks but eventually the organisms dissociate and the whole bizarre fertilization analogue begins again. *Geosiphon*, genus and species, originated—like so many others—by symbiogenesis. Both from the evolutionary and the developmental point of view *Geosiphon* is entirely a product of symbiotic interactions.

NITROGEN-FIXING NICHES

Other plant genera and species originated symbiogenetically, and for the usual reason: nitrogen starvation. Rice fields tend to be bright green. The green low-lying patches between the rice stalks look like algae. The stalks have no obvious stems, leaves, or roots. Careful examination shows that they are plants—what seem to be the flat thallus of an alga are leaves. These rice paddy dwellers are tiny water ferns. In pouches on the backs of their leaves live *Anabaena*, another species of filamentous cyanobacteria adept at the fixation of nitrogen. *Anabaena* glides into the immature leaf cavity which then grows by closing around the opening. Six species of *Azolla* have been named and studied. Special cells of *Azolla* act as "racks" to hold *Anabaena* in place.

Cycads are familiar to anyone who visits Miami Beach and other Florida resort cities in winter. They look like giant pineapples plopped on the ground. Many decorate hotel lobbies in huge round pots. They are native to the tropics worldwide and they are especially conspicuous in Cuba and South Africa. Cycads form a "higher taxon," a phylum or division of plants with dozens of species. All cycads are symbiotic with nitrogen-fixing cyanobacteria. None exist in nature without their captive bacterial genomes. We suspect that the cyanobacterial association began the cycad species proliferation. The cyanobacteria are not found in the seeds but they enter the seedling within twenty-four hours after it germinates. Tree-sized cycads form little specialized vertical organs in the surrounding soil, called "coralloid roots." Cut into one and you'll see a fully integrated little greenish blue ring that to the untrained eye is indistinguishable from the cells in the rest of the plant. This is where the prokaryotic cells of nitrogen-fixing cyanobacteria (*Nostoc* again) have been acquired and integrated into the cycad's root cells.

The plethora of Cretaceous leguminous plants, trees, herbs, vines, and the like, members of the spectacularly successful order Leguminoseae, evolved by symbiogenesis. The story is too long, circuitous, and technical to be told here. Just remember the next time you eat a pea or a bean, that one reason this nitrogen-rich food is so good for you is that the plant from which it was taken harbors a bacterial genome in its roots. The intimacy of the relationship between the legume (by itself totally incapable of using nitrogen from the air) and the bacterium (which by itself does not fix nitrogen either) is remarkable. These partners are integrated on the genetic, gene product, gene, metabolic, and behavioral levels simultaneously. The association between the soil rhizobial bacteria and the little roothairs of the bean plant that welcome their swimming, single-file invasion is at least 100 million years old. The productive symbiosis is probably behind the proliferation and speciation of this marvelous family of flowering plants. The bacterial nitrogen fixers are not photosynthesizers. Indeed, in legumes, they depend on sugars and other photosynthate from the plants.

In the case of the legumes, the details of bacterium acquisition and cyclical integration are fairly well known because entrepreneurs seek to imitate nature. Agricultural companies want to teach corn and wheat plants (members of the grass family) to capture the genomes of the bacteria, rhizobia, to produce grain that is richer in protein and cheaper than other nitrogen sources. So far it has not worked. The subtleties of the legume-rhizobia integration story are many. If these entrepreneurs were successful a *Zea* (corn), a *Hordeum* (barley), or a *Triticum* (wheat) plant that fixed nitrogen would certainly be endowed with a new species name: *Zea azotogenica*, *Hordeum nitrogenicum* or perhaps *Triticum nutricium*. Again we would be seeing the origin of species not by random mutation but by genome acquisition.

CHAPTER 12

CHROMOSOME DANCE: THE FISSION THEORY

*N*o one is more acutely aware of the importance of chromosomes than prospective parents awaiting the results of amniocentesis. They want to be reassured that the cells of their fetus have no more or no fewer than forty-six chromosomes, neatly aligned in twenty-three matching pairs. If they yearn for a boy child then only the last pair (the XY) may vary. The large X will not match the small Y. It is the Y chromosome, which carries very few genes (mainly for sperm formation and swimming), that determines the gender of the boy child. All the cells of healthy girls invariably have paired chromosomes, including the X chromosomes, of which girls and women have two. The amniocentesis procedure essentially is a karyotype analysis. An axiom of mammalian biology is that all members of the same species tend to have identical, or at least extremely similar, karyotypes.

The karyotype is defined as the total number and the morphology of the entire chromosome set of an animal or plant. Karyotype preparations come from body cells in the process of mitotic division, from which permanent preparations may be made and stored. The cells are fixed (treated with chemicals that maintain discernible structure) and stained with chromosome dyes that permit the details of the chromosomes to be tabulated and photographed. The chromosome pairs are then numbered and aligned by size from large to small. Either digital images or actual photographs of the chromosomes are interpreted. They must be of high enough quality to assure the karyotype can be read. Often a new human life is at stake.

Karyotype details have been collected for over a thousand mammals and hundreds of plants since the early twentieth century. Since karyotypes generally are species-specific, karyological analysis is a useful aid in species identification.

If Darwinian gradualism explains the origins of animal and plant species, it follows that closely related species should have similar karyotypes. They don't. Take the small hardy Asian deer, species of the genus *Munjiacus*. These deer look very similar to each other and, when tested, they apparently are capable of mating to produce offspring. The seven or eight species of *Munjiacus* range from west to east Asia, over an area of several million square kilometers. Although certainly all the data are not in, their diploid (body cell) chromosome numbers vary from a low of just three pairs to a high of twenty-three.

Neil Todd, as a fresh Ph.D. in zoology from Harvard University, in the early 1970s, developed a theory of karyotypes to correlate the evolution of chromosomes with the evolutionary history of mammals. He named the concept *Karyotypic Fission Theory*, two commentators remarked, to "call attention to his implicit rejection of Darwinian gradualism in chromosomal evolution." Todd's original analysis applied to three groups of animals: to canid carnivores, to artiodactyls (goats, sheep, deer, and other two-toed ungulates), and (with a Boston University graduate student, John Giusto) to old world monkeys and apes.

His papers were ignored or rejected by mainstream evolutionists. M. J. D. White, in his 1973 book *Animal Cytology and Evolution*, dismissed Todd's central thesis with the following remark: "To suppose that all of the chromosomes of a karyotype would undergo this process [fissioning at the centromere-kinetochore] simultaneously is equivalent to a belief in miracles, which has no place in science" (p. 401). Not surprisingly, few zoologists and evolutionists were inspired to take Todd's work seriously. Unlike Professor White, Todd had no laboratory of his own for the study of chromosomes. He was not a senior professor or the author of a widely cited textbook. In fact Neil Todd was and is a numismatist and genealogist, with impressive expertise both in seventeenth- and nineteenth-century Irish tavern tokens in cat migration and population genetics.

Zoologists and evolutionists continued to work under the assumptions, widely touted by White and other neodarwinists, that chromosome numbers were extremely similar in related species of mammals and that the exceptions Todd had noted were trivial. The mantra, widely repeated, was that mammalian chromosome numbers were high in ancestral populations and gradually decreased over time. The fusion of chromosomes, as happens in children with Down's syndrome, occurred unpredictably and accounted for chromosome changes. The word was that Todd, a believer in miracles, should be ignored. With the exception of a few sporadic papers the entire karyotypic fission story lay dormant until 1998.

The story has resumed for several reasons that are relevant to the thesis of this book. Karyotypic fissioning itself is not a symbiogenetic process, but we suggest it is related ultimately to the symbiogenetic origin of the centromere-kinetochore of the chromosomes (Todd, 2001).

Because of the extraordinary growth of the field of cell biology and the study of chromosomes, the talent of a young woman biologist, Robin Kolnicki, who analyzed the chromosome data for lemurs, and Todd's own rekindled interest in his great idea, the situation has radically changed. Todd's vision is the best and most parsimonious explanation for mammalian karyotypes and chromosome change through time. Kolnicki's mustering of much evidence from a different field of life sciences shows why and how, without miracles, Todd's ideas about adaptive radiations turn out to be correct. "Adaptive radiation" refers to proliferation in species diversity at a given time in the fossil record, which is here correlated with dramatic changes in mammalian chromosome numbers. New young investigators, less likely to accept unthinkingly the assumption of gradual tiny steps by which evolution is supposed to proceed, may more and more apply the Todd-Kolnicki view (now called *Kinetochore Reproduction–Karyotypic Fission Theory*) to bats, rodents, other mammals, and even to birds and reptiles. The origins of new mammalian species correlate with karyotypic change in just the saltatory, discontinuous way that Todd envisioned. Here we can only cursorily review the basic ideas. The scientific literature is still limited and manageable because it was only recently jump-started. Like the other aspects of neodarwinian demise related to new evidence, the debate has become lively.

When karyotype analysis, spurred by the chromosome theory of heredity, began in the 1930s, many scientists were surprised to find that similar animals did not necessarily have similar karyotypes. Diploid numbers in mammals vary from a single pair of chromosomes to forty-six pairs. Todd's fission theory postulates an ancestral mammal with a diploid number of fourteen "mediocentric" chromosomes (the kinetochore-centromere lies near or at the center of the chromosome). Complete fissioning of all the chromosomes at once is hypothesized to underlie spurts of speciation (Figure 12.1).

The fissioning event, in populations with all large mediocentric chromosomes, will generate a full set of smaller, acrocentric chromosomes (with the centromere near one end) and a diploid number doubled relative to the original. The diploid number in the thirty-two species of lemurs ranges from twenty chromosomes to seventy.

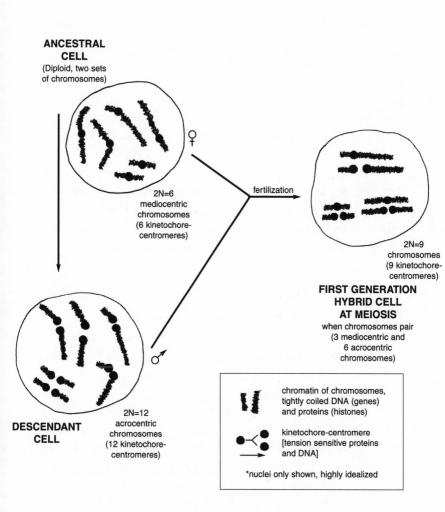

FIGURE 12.1   Karyotypic Fission (Kinetochore-Centromere Reproduction) Theory

Lemurs, a suborder of primates native to Madagascar, group into five families: Lepilemurids, Daubentonids, Lemurids (or Eulemurs), Cheirogalids, and Indrids. A minimum of four evolutionary steps, beginning with the hypothetical ancestor (probably resembling today's *Lepilemur ruficandatus*) with a diploid number of twenty, in principle, generates karyotypes in all living lemur species. A primary fissioning event led to a diploid number range of twenty to thirty-eight chromosomes. A second event explains lemurids (forty-four to sixty-two chromosomes) and Cheirogalids. In the ancestors of Indrids a different secondary fissioning occurred, as well as a later separate fissioning in the Lepilemurids. The details need to be filled in, but fission theory elegantly reduces the need for hundreds of ad hoc explanations to a very few fissioning events followed by smaller changes in chromosomes in limited groups of descendants.

KINETOCHORE REPRODUCTION AS BACTERIAL LEGACY

Skip this section if the scientific terms are burdensome. Suffice it to know that the tendency of chromosomes to reproduce, break, regroup, and change on their own schedules probably reflects their past, free-living lives. The behaviors of these gene-bearing structures profoundly influence us, the animals and plants whose growing cells cannot do without them. This section discusses how the *Kinetochore Reproduction Theory* explains, in new biological terms and in one fell swoop, Todd's thirty-year-old *Karyotypic Fission Theory* by a single plausible event and its consequences.

Kinetochore-centromeres are the places of attachment of each chromosome to the mitotic spindle, the microtubule tracks upon which the chromosomes ride in order to get to the far sides of the dividing cell. They are now known to be double structures of protein (the kinetochore) that attach to the mitotic spindle microtubules and keep the DNA (the centromere) continuous with the rest of the genes in a single package. Each time a chromosome reproduces, the kinetochore-centromere protein-DNA package reproduces as well. The chromosome-kinetochore-centromere reproduction occurs each time the cell itself reproduces.

The single plausible event is an extra round of kinetochore-centromere reproduction followed by chromosome breakage between the kinetochore pairs. This extra round and breakage occurs in all the chromosomes at once. For example, if it were to occur in your sperm-making cells it would happen in all of your twenty-three pairs of chromosomes, which have their centromeres located mediocentrically. This single plausible event would generate a new cell now with forty-six pairs of shorter chromosomes. In each new chromosome the kinetochore-centromere would now be at the end. In the jargon, the karyotypic fission event converted forty-six mediocentrics to ninety-two telocentrics with no significant change in gene sequence (order of DNA base pairs), no change in gene dosage (the ratios of all genes to each other stay the same), no change in DNA quantity (the distribution of genes into more linkage groups occurred without any alteration in the total quantity of DNA), nor any other change in the animal itself. Chromosome re-arrangements and sorting out of groups of chromosomes after the great fissioning is inevitable because when you mate and your fissioned sperm fertilizes an egg, it is a foregone conclusion that your sweetheart's egg will have the normal twenty-three long mediocentric chromosomes (not forty-six short ones like yours). Why? Because she and her cells will be the number typical of our species.

Nature has tried this experiment many times: The fissioned and the nonfissioned animals are fully fertile until chromosome rearrangements (pericentric inversions and the like, beyond the scope of this discussion) occur. The kinetochore reproduction explanation of the fissioning depends crucially on changes (acceleration) in the timing of new kinetochore-centromere production. These changes in timing, a delay in chromosome reproduction relative to reproduction of the kinetochore-centromere, underlie much mammalian species diversification. In certain groups of mammals these changes have been shown in detail to explain the taxa: carnivores, even-toed (artiodactyl) mammals (such as sheep, goats, and deer), lemurs and old world monkeys, and apes (chimps, orangutans, gorillas, cercopithecid monkeys, baboons and the like).

Kinetochore-centromeres are distinguishable sites on chromosomes, pinched-in regions where spindle fibers (microtubules) attach (Figure 12.1). Kinetochore-centromeres belong to a class of structures called "microtubule-organizing centers" (MTOCs) because they are seen to "capture" microtubules and attach them to the rest of the mitotic spindle. New kinetochore-centromere production generally occurs between mitoses when all the other chromosomal DNA is also reproducing. An extra round of synthesis of kinetochore-centromere DNA relative to the rest of the chromosome is all that is needed to start the karyotypic fissioning ball rolling down the hill of new mammalian species.

The new analysis of Todd's theory incorporates newly discovered biochemical facts that correlate the composition of kinetochore-centromeres with their behavior as the movement of mitotic cell division occurs. Proteins that comprise the kinetochore are sensitive to mechanostimulation—to pulling. The addition and removal of specific chemicals (phosphate ions) and the presence and activities of microtubules are caused by external stimulation. When tickled or prodded the kinetochore-centromeres respond. They tend to do their thing, which is to reproduce. Kolnicki points out that the idea that this part of the chromosome reproduces, occasionally out of sync with the rest of the chromosomes, is consistent with our ideas of the origin of nucleated cells in general. She wrote, "Endocellular symbionts tend to reproduce out of synchrony from their hosts even in co-evolved eubacterial symbiotic associations like those of mitochondria and plastids. Kinetochore reproduction theory (karyotypic fission theory) where centromere 'splitting' is understood as rapid centromeric residual reproduction of a once-foreign (once spirochete eubacterial) genome is entirely consistent with a 'symbiogenetic' rather than a 'direct filiation' concept of eukaryotic cell evolution."

Of course we agree. We add the saltational, discontinuous Todd-Kolnicki idea of speciation to ideas such as symbiogenesis and Williamson's larval transfer. These are some of the ways that evolutionary novelty is created. Species originate by inheritance of acquired genomes—and their reproductive rebellions within host genomes. So do higher taxa such as genera, families, and the like. The process is not gradual—nor, if we consider these symbiotic "aftershocks" in the fissioning of mammalian chromosomes, is it one from which animals are exempt.

CHAPTER 13

DARWIN REVISITED: SPECIES IN THE EVOLUTIONARY DIALOGUE

*I*n this book we have questioned the adequacy of the popular modern evolutionist explanation of the origins of new, heritable features of life and the evolution of new species and higher, more inclusive taxa. The reliance on accumulation of random mutations in DNA is not so much "wrong" as oversimplified and incomplete: It misses the symbiotic forest for the genetic trees. The neodarwinists' inventive literature and valiant attempts to unite the genetic stability in the unblended mixture of Gregor Mendel's factors to the gradual evolutionary change promoted by Darwin's natural selection were as brilliant as they were incorrect. The hegemony of R. A. Fisher, J. B. S. Haldane, and Sewall Wright is gone forever, and their latter-day saints—Richard Dawkins and J. Maynard Smith, or at least their students—will have to learn something about chemistry, microbiology, molecular biology, paleontology, and the air. Better-informed scientists have reinstated, in the light of new knowledge, the Darwinian, not neodarwinian, concepts of evolution as the organizing principles for the understanding of life. We suggest that at least some of Jean Baptiste de Lamarck's "acquired characteristics" that sensitively respond to the exigencies of the environment are foreign genomes. Tiny masters of metabolism and movement are often ready and willing to associate with larger forms when environmental pressures encourage togetherness. Evolution's menagerie is far more responsive to immediate environmental forces than the "random mutation" contingent would have us believe. The branches of the evolutionary tree fork, but they also fuse. Genomes integrate; the mergers persist past the point of no return. Evolution is irreversible. Even the evolution of the eye has a symbiotic, as well as a random mutation, component to its history. And like all evolutionary novelty, it must be seen as the several results of protracted biological, geological, and historical paths. Indeed, the evolution of vertebrate organs as complex as eyes has always fascinated everyone concerned with the origin of genuine innovation. How does an eyeless animal "grow an eye"? (Samuel Butler remarked that "We don't remember when we first grew an eye.") Anthropocentric writers with a proclivity for the miraculous and a commitment to divine intervention tend to attribute historical appearances like eyes, wings, and speech to "irreducible complexity" (as, for example, Michael Behe does in his book, *Darwin's Black Box*) or "ingenious design" (in the tradition of William Paley who used the functional organs of animals as proof for the existence of God). Here we feel no need for supernatural hypotheses. Rather, we insist that today, more than ever, it is the growing scientific understanding of how new traits appear, ones even as complex as the vertebrate eye, that has triumphed. What is the news?

The fundamental idea is that inherited characteristics of extraordinary importance to natural selection, such as fruits, eyes, wings, or speech, always enjoy a long history of precedence. Awareness of the microbial antecedents and eukaryotic genetics and development converts what may seem divine intervention to evolution as usual. The evolutionary appearance of eyes is no exception. The capacity to specifically respond to visible light appears in many molecules common to all cells. The photosensitive ability of the human (and all vertebrate) retinal tissue is a property of the rods and cones of retinal cells. Chemically, these cells harbor rhodopsin, a purplish light-sensitive protein-pigment complex. The opsin portion of rhodopsin is a protein that varies from one organism to another but shares common features and is present in a huge range of animals, some bacteria, and protists. The actual light reactions occur in the second, smaller portion of the molecule: the retinal. The name, "*rhodo*," Greek for purple, refers to the retinal part, and *opsin*, where the "op" is the same as in optician or optical, refers to "eye." Loosely then, the apt name of the active molecule is purple eye protein. The small molecular portion that receives the light directly dates back in the history of life to a time long before animals and plants. Rhodopsin is conspicuously active in the entire group of archaebacteria called halophils. These salt-loving prokaryotes bathe in sunlight; they use their light-sensitive rhodopsin to generate energy in the form of the ubiquitous molecule ATP. The fundamentals of photosensitivity are already well developed in the bacterial world.

The rods (used mainly for night vision) and cones (mainly for daylight) are elongate membrane-rich cells that line the retina in the eyes of animals. In our own eyes these cells bear the standard [9(3)+0] microtubular kinetosomes that underlay the microtubular shaft, the [9(2)+2] axoneme that we of course have studied in the context of the origin of intracellular motility. The rod is simply a greatly overgrown parallel infolding of the undulipodial (cilia) membrane. The cone is the same, although the folded membranes are in tapered stacks, rather than in parallel stacks, giving the entire cell a cone-like shape. If we concur that the entire undulipodial system derives from spirochete symbionts then, yes, symbiosis is also a prerequisite to the evolution of the eye. What has been shown by the computer-generated natural selective pattern of Richard Dawkins and extensive study of the literature by Ernst Mayr and his colleagues is that any population of organisms with surface photosensitivity will tend to refine and expand this capacity in the direction of an organ that is protected, that can focus and move and even form an image. According to Mayr, eyes evolved in animal lineages at least forty times. The most conspicuous and best-studied examples are the simple eyes of vertebrates and mollusks (like squid and clams) and the compound eyes of insects. Not included in their list is the evolution of a cameralike eye in two genera of single-celled protists (the erythrodinid dinomastigotes) where the entire cell has become a functional modified eye; indeed *Erythrodiniopsis* is entirely analogous to a wide-angle lens camera. The single cell whose plastid and other pigment-coated membranes form the retinal-equivalent sits attached as it waits for its microbial prey to cast a shadow overhead. As in any endeavor that seeks the origin of a complex, and superficially perfect, trait, one needs to study, in Darwin's words, "the oddities and peculiarities" of its antecedents.

The language of evolutionary change is neither mathematics nor computer-generated morphology. Certainly it is not statistics. Rather, natural history, ecology, genetics, and metabolism must be supplemented with accurate knowledge of microbes. Microbial physiology, ecology, and protistology are essential to understand the evolutionary process. The behavior of microbes within their own populations and in their interactions with others determined life's winding, expanding evolutionary course. The living subvisible world ultimately underlies the behavior, development, ecology, and evolution of the much larger world of which we are a part and with which we co-evolved. While some may feel belittled by this perspective of evolution punctuated and driven forward by microbial mergers, we believe, echoing Darwin, that there is grandeur, too, in this view of life. Numberless forms and variation come not just gradually and at random, but suddenly and forcefully, by the co-opting of strangers, the involvement and infolding of others—viral, bacterial, and eukaryotic—into ever more complex and miscegenous genomes. The acquisition of the reproducing other, of the microbe and its genome, is no mere sideshow. Attraction, merger, fusion, incorporation, cohabitation, recombination—both permanent and cyclical—and other forbidden couplings, are the main sources of Darwin's missing variation. Sensitivity, co-optation, merger, acquisition, fusion, accommodation, perseverance and other capabilities of the microbes are not at all irrelevant to the evolutionary process. Far from it. The incorporation and integration of "foreign" genomes, bacterial and other, led to significant, useful heritable variation. The acquiring of genomes has been central to the evolutionary processes throughout the long and circuitous history of life.

Indeed, as Wallin wrote in 1927, "It is a rather startling proposal that bacteria, the organisms which are popularly associated with disease, may represent the fundamental causative factor in the origins of species." We agree.

GLOSSARY

Akaryomastigont   Intracellular organellar complex found in many mastigotes (undulipodiated cells). The mastigont system includes the axoneme shaft and surrounding membrane of the undulipodia, the underlying kinetosome from which the undulipodium arises, and associated fibers, which may include costa, parabasal bodies, and axostyle. An akaryomastigont system, unlike the karyomastigont system, is not connected to the nucleus via a nuclear connector, or rhizoplast. (Figure 9.3)

Amitochondriate   An anaerobic eukaryotic organism that lacks mitochondria, either primarily (its eukaryotic ancestors had never acquired mitochondria to begin with) or secondarily (it lost the mitochondria that were present in its ancestors).

Archaeprotista   Phylum within kingdom Protoctista comprised of primarily amitochondriate anaerobic or sometimes microaerophilic protists, including amebas and mastigotes (undulipodiated single-celled organisms) all without mitochondria.

Archamoebae   One of the three classes (Archamoebae, Metamonada, and Parabasalida) of the phylum Archaeprotista. Archamoebae include free-living freshwater and marine microaerophilic amitochondriate amebas that lack undulipodia and the anaerobic mastigamoebae amebas that do bear undulipodia at some stage in their life history.

Association   Regular ecological relationship of two or more different kinds of organisms. See "symbiosis."

Bacteria   Members of kingdom Bacteria (also called Monera or Prokaryotae), one of the five kingdoms of life. All bacteria and only bacteria are prokaryotes; they lack a membrane-bounded nucleus. Sometimes divided into two subkingdoms, the eubacteria and the archaebacteria.

Biosphere   All of the places on the Earth's surface, from the upper limits in the atmosphere to the lowest depths in the ocean, where life exists.

Biotic potential   The number of organisms that can be produced in a single generation, or unit of time, which is characteristic of a given species, measured in maximum number of offspring per generation, maximum number of spores produced per year, or equivalent terms. It illustrates the tendencies of organisms to increase exponentially when their conditions for material growth are satisfied.

Centromere   Structure attaching chromosomes to microtubules of mitotic spindle. Microtubule-capturing center located on chromosomes. Centromeric connections to the spindle are required for chromatid segregation. The centromere, as a region of the chromosome deduced from genetic behavior, is sometimes distinguished from the kinetochore as a structure observable in the electron microscope. Some authors consider centromeres synonymous with kinetochores.

Chemoautotrophy   A metabolic mode of nutrition by which an organism obtains energy by the oxidation of inorganic substrates such as sulfur, nitrogen, or iron and makes its cell carbon from CO_2 (carbon dioxide).

Chimera   An organism formed by the merger of two or more genetic types, formed by symbiosis, abnormal chromosome segregation, or artificial grafting; usually used in reference to microorganisms and to plants, rarely to an animal; a mosaic.

Cilium   Undulipodium. Organelle of motility that protrudes from the cell comprised of an axoneme covered by the plasma membrane. The term is used to refer to undulipodia of ciliates and of animal tissue cells. Composed of the [9(2)+2] microtubular shaft, axoneme.

Classification   A process of establishing, defining, and ranking taxa within hierarchical series of groups, whether artificial or natural. A scheme of such a hierarchical series of groups or taxa.

Consortium   A group of individuals of different species, typically of different phyla, living in close association; in microbiology, a physical association between the cells of two or more different types of microorganism that is more or less stable over time.

Epibiont   Ecological term describing the topology of association of partners in which one organism (the epibiont) lives on the surface of another (different kind of) organism.

Epiphyte   A plant growing on another plant (the phorophyte) for support or anchorage rather than for water supply or nutrients. Any organism living on the surface of a plant.

Eukaryotes   Organisms made of nucleated cells, cells with at least one membrane-bounded nucleus that undergo some form of mitosis. Eukaryotes (kingdoms Protoctista, Animalia, Fungi, and Plantae) all derive from and include the protoctists; kingdom Bacteria (the prokaryotes) lack nucleated cells.

Euplotidium   A ciliate protist (Phylum Ciliophora) with projectile bacteria.

Euprymna scolopes   Commonly called the Hawaiian bobtail squid, this cephalopod mollusk harbors the symbiotic bioluminescent bacteria *Vibrio fischeri.* Symbiotic evolution has led to the squid's complex light-emitting organ, in which the bacteria are contained.

Evolution   Any cumulative change in the characteristics of organisms or populations from generation to generation; descent with modification. All the changes that have transformed life on Earth from its earliest beginnings to the diversity that characterize it today.

Evolutionary biologist   A biologist who studies the evolution of life, either in general or of a specific organism or organisms, whether it be on an ecosystem, ecological, community, population, individual, cellular, or molecular level. One who studies the integrated science of evolution, ecology, behavior, and systematics.

Exponential growth   Growth that is unimpeded by checks, so that the larger the population the faster it grows. The geometric increase of a population as it grows in an ideal, unlimited environment.

Flagellum   Bacterial flagellum: prokaryotic extracellular structure composed of homogeneous protein polymers, members of a class of proteins called flagellins; moves by rotation at the base; relatively rigid rod driven by a rotary motor embedded in the cell membrane that is intrinsically nonmotile and sometimes sheathed. Term sometimes incorrectly used to refer to the eukaryotic undulipodium, the intrinsically motile intracellular structure used for locomotion and feeding in eukaryotes composed of a standard arrangement of nine doublet microtubules and two central microtubules composed of tubulin, dynein, and approximately 200 other proteins, none of them flagellin. No flagellum (but every undulipodium) is underlain by a kinetosome.

Geological Time Scale   A time scale established by geologists that reflects changes in geological history. The major divisions are eons: Hadean, Archean, Proterozoic, and Phanerozoic. Traditional paleontology tends to focus only on macroscopic events from the last eon, the Phanerozoic. (See Table 9.1)

Germs Nonscientific term used to refer to microbes that may or may not cause illness. Term used to refer to microbes as something dirty that people want to avoid. Contagious disease—associated "enemies." Does not distinguish viruses, bacteria, protists, prions, etc.

Golgi apparatus Dictyosome; golgi body. Organelle system of membranes of nearly all eukaryotic cells visible with the electron microscope as a structure of flattened vesicles, or cisternae, often stacked in parallel arrays; involved in elaboration, storage, and secretion of products of cell synthesis.

Gunnera The only angiosperm (flowering plant) group that forms regular symbiotic relationships with cyanobacteria. There are forty-five species within the genus *Gunnera*, all of which are associated with symbiotic nitrogen-fixing *Nostoc* cyanobacteria. The largest species, *G. manicata*, is known by the common name "poor man's umbrella."

Hermissenda A genus of nudibranch mollusks (sea slugs) that live in symbiosis with tubullarian cnidarians, sessile marine animals that possess stingers called nematocysts. (Jellyfish, corals, and sea anemones are all also within the phylum Coelenterta.) The nudibranch, able to feed on the tubullarian hydroids without getting stung, stores the stingers in specialized organs to use for its own defense.

Heterotermes tenuis Subterranean wood-eating termite belonging to the family Rhinotermitidae. Widespread in the Americas. Known to harbor symbiotic protists, such as *Holomastigotoides, Pseudotrichonympha,* and *Spirotrichonympha* in its hindgut.

Hybrid Offspring of a cross between genetically dissimilar individuals; often restricted in taxonomy to the offspring of interspecific crosses. A community comprising taxa derived from two or more separate and distinct communities. In nucleic acid metabolism, a double-stranded polynucleotide, one strand being a DNA and the other an RNA.

Karyomastigont Intracellular organellar complex found in many mastigotes (undulipodiated cells). Organelles associated with undulipodia, the mastigont system may include the kinetids with their undulipodia, undulating membrane, costa, parabasal bodies, and axostyle, all connected to the nucleus via a nuclear connector, such as a rhizoplast. Compare with "akaryomastigont."

Karyotype   Total chromosome complement of an animal, plant, fungus, or protoctist as seen in fixed and stained preparations of condensed chromosomes using the light microscope; karyotyping is a fixation and staining procedure used to determine characteristic morphology and the number of chromosomes for a species.

Kinetochore   Microtubule-organizing center usually located at a constricted region of a chromosome that holds chromatids together. Kinetochores, morphologically visible manifestations of centromeres, are the site of attachment of microtubules forming the spindle fibers during nuclear division (mitosis and meiosis). Centromeres are deduced from genetic behavior whereas kinetochores are directly visible by electron microscopy.

Kinetosome   An intracellular organelle, not membrane-bounded, characteristic of all undulipodiated cells. Microtubular structures necessary for the formation of all undulipodia, kinetosomes differ from centrioles in that from them extend the shaft, or axoneme. Their microtubules are organized in the [9(3) + 0] array; all undulipodia are underlain by kinetosomes. These basal organelles, often called "basal bodies" because they generate the axoneme, are necessary for the formation of all undulipodia; kinetosomes differ from centrioles in that from them extend [9(2)+2] axonemes. The term "kinetosome," because of its precision, is preferable to "basal body."

Larva   A free-living, sexually immature form in some animal life histories that may differ from the adult in morphology, nutrition, and habitat.

Mendel, Gregor Johann   Austrian monk (1822–1884), the father of Mendelian genetics. His static principles of genetics: dominance in a heterozygote (one allele may conceal the presence of another) and the principle of segregation in a heterozygote (two different alleles segregate from each other during the formation of gametes) combined with Darwin's "change-through-time" idea formed the population genetics (neodarwinist) view of evolution.

Metabolic mode   The type of metabolism an organism has, based on whether it uses light or chemicals for energy, whether it uses inorganic or organic compounds for its source of electrons, and whether it uses CO_2 or eats food for its carbon source.

Metabolism   The sum of the chemical and physical processes that occur in all living organisms and involve incessant replacement of their chemical constituents.

Metamonada   One of the three classes (Archamoebae, Metamonada, and Parabasalida) of the phylum Archaeprotista. Metamonads are anaerobic amitochondriate protists that have karyomastigonts; their nuclei are attached to their undulipodia via nuclear connectors. *Giardia* is a disease-causing free-living metamonad. Many are symbiotic in the intestines of insects.

Microbes (Microorganisms)   Beings (usually bacteria, protoctists, and fungi) that are best seen through a microscope.

Mollusk (Mollusc)   A diverse phylum of animals, comprised of seven classes of marine, freshwater, and land organisms. Most mollusks have an internal or external shell, a muscular foot, and an unsegmented, soft body. Includes clams, chambered nautilus, squid, octopus, snails, slugs, and nudibranchs.

Monophyly   The condition of a trait or a group of organisms considered to have evolved directly from a single common ancestor. Sister taxa are said to be monophyletic.

Monotheism   The religious belief in one God, who is supernatural and all-powerful. Judaism, Christianity, and Islam are all examples of monotheistic faiths.

Natural selection   The process resulting from more organisms being produced than can ever survive. Those that survive to have offspring are said to be naturally selected.

Nematocyst   Cnidocyst. Modified cell with a capsule containing a threadlike stinger used for defense, anchoring, or capturing prey; some contain poisonous or paralyzing substances (for example, in all coelenterates and ctenophores); analogous organelles found in some dinomastigotes and some karyorelictid and suctorian ciliates.

Neodarwinism   Belief that random mutation is the major source of evolutionary change, that natural selection acts upon. Derived from fusion of Mendel's genetics with Darwin's 'descent with modification.'

New Synthesis (Modern Synthesis)   A comprehensive theory of evolution emphasizing natural selection, gradualism, and populations as the fundamental units of evolutionary change. Neodarwinism.

Nuclear connector (rhizoplast)   Cross-banded microfibrillar ribbon extending from the bases of kinetosomes and directed toward the nucleus or to cyoplasmic microtubule-organizing centers. Cell orgenellar system.

Nucleoid   DNA-containing structure of prokaryotes, not bounded by the nuclear membrane; does not contain nuclear pores. Visible by electron microscopy.

Nucleus   Membrane-bounded, spherical, DNA-containing organelle, universal in protoctist, animal, plant, and fungal cells. Chromatin (DNA, protein) organized into chromosomes; site of DNA synthesis and RNA transcription. Defining organelle of eukaryotes.

Nudibranch   A gastropod mollusk without a shell, a sea slug. (Other gastropod mollusks include whelks and land snails.)

Olenus, olenids   A taxon of lower Paleozoic trilobites that lived under oxygen-poor, sulfur-rich sea floor conditions which were widespread in the region where Scandinavia is today. These areas did not support any other macroscopic life forms, and it is thought that the olenid trilobites lived in this harsh environment because of chemoautotrophic bacterial symbionts.

Parabasalida   One of the three classes (Archamoebae, Metamonada, and Parabasalida) of the phylum Archaeprotista. Parabasalids are symbiotic protists that live in the intestines of insects. They apparently digest cellulose, from which they derive sugars. A parabasalid bears at least four undulipodia, an axostyle, and conspicuous kinetosomes. The two orders of Parabasalida are Trichomonadida and Hypermastigida.

Photoautotrophy   Mode of nutrition in which light provides the source of energy. An obligately photoautotrophic organism uses light energy to synthesize cell material from inorganic compounds (carbon dioxide, nitrogen salts).

Photosynthesis   A process by which chemical energy is generated from sunlight found in certain bacteria, algae, and most plants.

Phylogeny   Hypothesized sequence of ancestry of groups of organisms as depicted in their evolutionary trees. Compare with "systematics" and "taxonomy."

Prokaryotes   Bacterium; member of the kingdom Bacteria (kingdom Monera or Prokaryotae); cell or organism composed of cells with nucleoids, lacking a membrane-bounded nucleus. (Example shown in Figure 9.2.)

Protists   The single-celled or very-few-celled and therefore microscopic members of kingdom Protoctista.

Protoctista   One of the five kingdoms into which all living organisms are classified. Protoctists are eukaryotic, nucleated microorganisms (the single-celled protists) and their direct multicellular descendants. The kingdom includes all eukaryotic organisms with the exception of animals, plants, and fungi; for example, all algae, slime molds, amebas, slime nets, water molds, and foraminifera are among the estimated 250,000 extant species in about fifty major groups. (Examples shown in Figures 7.4, 7.5, 9.1, 9.3, and 9.4.)

Snyderella   Multinucleate termite symbiont protist, a calonymphid (Figures 7.4 and 9.4).

Species   A group of organisms, minerals, or other entities formally recognized as distinct from other groups. A taxon of the rank of species; in the hierarchy of biological classification, the category below genus; the basic unit of biological classification. In this book we suggest that organisms with the same kinds and numbers of integrated genomes in common are members of the same species.

Staurojoenina   Hypermastigote termite symbiont protist (Figure 7.5).

Symbiosis   Prolonged physical association between two or more "differently named" organisms, generally from two or more different species. Levels of partner integration in symbiosis may vary in intimacy; integration may be behavioral, metabolic, of gene products, or genic. Members of a symbiosis are termed symbionts.

Systematics   That subfield of evolutionary science that deals with naming, classifying, and grouping organisms on the basis of their evolutionary relationships. Compare with "phylogeny" and "taxonomy."

Taxon (plural taxa)   A group of similar organisms, such as kingdom (most inclusive), phylum, family, genus, or species (least inclusive taxon).

Taxonomy   The theory and practice of describing, naming, and classifying organisms. Compare with "systematics" and "phylogeny."

Thiodendron   Dubinina's spirochete in association with a sulfate-reducing bacterium. A marine consortium first published as a single bacterium.

Trilobite Extinct group of animals that lived during the Cambrian. Segmented, hard-shelled arthropod. Common fossil.

Undulipodium Cell-membrane-covered motility organelle sometimes showing feeding or sensory functions; composed of at least 200 proteins. [9(2)+2] microtubular axoneme covered by plasma (cell) membrane; limited to eukaryotic cells. Includes cilia and sperm tails. Each undulipodium invariably develops from its kinetosome, a ninefold symmetrical microtubular structure at the base. Contrasts in many ways with the prokaryotic motility organelle or flagellum, a rigid structure composed of a single protein. Undulipodia in the cell biological literature are often referred to by the outmoded term "flagellum" or "euflagellum." Synonym: eukaryotic flagellum.

Zoocentrism Preoccupation with animals, including humans, as if animals were the main organisms in existence, and/or the only ones worthy of study. Great disregard for members of the other four kingdoms of life, dismissal of them as "lower" forms, ignores the major impact that these four kingdoms have upon members of the animal kingdom and Earth's ecosystems.

REFERENCES

Atsatt, Peter, 2003, The mycosome hypothesis: Fungi propagate within plastids of senescent plant tissue. *International Microbiology*, 6:1–25.

Behe, Michael, 1996, *Darwin's black box.* New York: The Free Press.

Brodo, I. M., S. D. Sharnoff, and S. Sharnoff, 2001, *Lichens of North America.* New Haven: Yale University Press.

Caldwell, D. E., 1999, Post-modern ecology—is the environment the organism? *Environmental Microbiology* 1:279–281.

Chapman, M. J., M. F. Dolan, and L. Margulis, 2000, Centrioles and kinetosomes: Form, function and evolution, *Quarterly Review of Biology* 75:409–429.

Diamond, Jared, 1999, *Guns, germs and steel: the fates of human societies.* New York: W. W. Norton.

Dolan, M., 2001, Speciation of termite gut protists: The role of bacterial symbionts, *International Microbiology* 4:203–218.

Dolan, M., H. Melnitsky, R. Kolnicki, R. and L. Margulis, 2002, Motility proteins and the origin of the nucleus, *Anatomical Record*, 268:290–301.

Fortey, Richard, 1998, *Life: An unauthorized biography.* Oxford, U.K.: Oxford University Press.

_____, 2000, Olenid trilobites: The oldest known chemoautotrophic symbionts? *Proceedings of the National Academy of Sciences* 97:6574–6578.

_____, 2000, *Trilobite! Eyewitness to evolution*. London: Flamingo, imprint of HarperCollins.

Freeman, S., and J. C. Herron, 2001, *Evolutionary analysis*, second edition. Englewood Cliffs, N.J.: Prentice-Hall.

Harold, Franklin, 2001, *The way of the cell*. New York: Oxford University Press.

Hoffman, Paul F., and Daniel P. Schrag, Snowball Earth. *Scientific American*, January 2000:68–75.

Kauffman, S., 2002, *Investigations*. New York: Oxford University Press, p. 45.

Keller, E. F., and E. A. Lloyd, 1992, *Key words in evolutionary biology*. Cambridge: Harvard University Press.

Keller, Laurent, 1999, *Levels of selection in evolution*. Princeton: Princeton University Press.

Khakhina, L. N., 1992, *Concepts of symbiogenesis: A historical and critical study of the research of Russian botanists*. New Haven: Yale University Press.

King, Jennifer M., Tom S. Hays, and R. Bruce Nicklas, 2000, Dynein is a transient kinetochore component whose binding is regulated by microtubule attachment, not tension. *Journal of Cell Biology* 151: 739–748.

Kolnicki, Robin, 1999, Karyotypic fission theory applied: Kinetochore reproduction and lemur evolution. *Symbiosis* 26:123–141.

_____, 2000, Kinetochore reproduction in animal evolution: Cell biological explanation of karyotypic fission theory. *Proceedings of the National Academy of Sciences* 97:9493–9497.

_____Kinetochore reproduction (=karyotypic fission) theory: Divergence of lemur taxa subsequent to simple chromosomal mutation. *Evolution* (in preparation).

Lapo, Andre, 1987, *Traces of bygone biospheres*. Moscow: Mir Publishers.

Lovelock, J. E., 1979, *Gaia: A new look at life on earth*. Oxford, U.K.: Oxford University Press.

_____1988, *The ages of Gaia: A biography of our living Earth*. New York: W. W. Norton.

_____, 2000, *Homage to Gaia: The life of an independent scientist*. Oxford, U.K.: Oxford University Press.

Lowman, Paul, 2002, *Exploring Space, Exploring Earth*. Cambridge, UK: Cambridge University Press.

Marcotrigiano, Michael, 1999, Chimeras and variegation: Patterns of deceit. *Horticultural Science* 32:773–784.

Margulis, Lynn, 1998, *Symbiotic planet. A new look at evolution.* New York: Basic Books.

_____, 1990, Words as battlecries: symbiogenesis and the new field of endocytobiology. *BioScience* 40:673–677.

Margulis, L., M. F. Dolan, and R. Guerrero, 2000, The chimeric eukaryote: Origin of the nucleus from the karyomastigont in amitochondriate protists. *Proceedings of the National Academy of Sciences* 97:6954–6959.

Margulis, L., and M. F. Dolan, 2002, *Early life,* second edition. Sudbury, Mass.: Jones and Bartlett.

Margulis, L., and L. Olendzenski, 2000, *Microcosmos* Vol. 1 (Cells and Reproduction) and Vol. 2 (Evolution and Diversity). Videos and accompanying booklets. Boston: Jones and Bartlett Publishers.

Margulis, Lynn, and Dorion Sagan, 1996, *Gaia to microcosm* Vol. 1. A Sciencewriter's Video. www.sciencewriters.org.

_____1997, *Microcosmos: Four billion years of microbial evolution.* Berkeley: University of California Press.

_____1997, *Slanted truths: Essays on Gaia, symbiosis and evolution.* New York: Copernicus Springer-Verlag.

_____2000, *What is life?* Berkeley: University of California Press.

Margulis, Lynn, and K. V. Schwartz, 1998, *Five kingdoms: An illustrated guide to the phyla of life on Earth,* third edition. New York: W. H. Freeman and Company.

Mayr, E., 1942, *Systematics and the origins of species.* Cambridge, Mass.: Harvard University Press.

_____, 1982, *Growth of biological thought.* Cambridge, Mass.: Harvard University Press.

_____, 2001, *What evolution is.* New York: Basic Books.

Mayr, E., and J. Diamond, 2001, *The birds of northern Melanasia: Speciation, ecology and biogeography.* Oxford, U.K., and New York: Oxford University Press.

Mehos, D. C., 1992, Appendix: Ivan E. Wallin and his theory of symbionticism. In Khakhina, L. N., *Concepts of symbiogenesis: A historical and critical study of Russian botanists.* New Haven: Yale University Press.

Miklos, G. L. G., 1993, Emergence of organizational complexities during metazoan evolution: perspectives from molecular biology, palaeontology and neo-Darwinism. Membership Association of Australasian Paleontologists, No. 15, pp. 7–41.

Morrison, Reg, 1999, *The spirit in the gene: Humanity's proud illusion and the laws of nature.* Ithaca, N.Y.: Cornell University Press.

Nicklas, Bruce, M. S. Campbell, S. C. Ward, and G. J. Gorbsky, 1998, Tension-sensitive kinetochore phosphorylation in vitro. *Journal of Cell Science* 111:3189–3196.

Oehler, Stefan, and Kostas Bourtzis, 2000, First international *Wolbachia* conference: Wolbachia 2000. *Symbiosis* 29:151–161.

Olendzenski, Lorraine, Lynn Margulis, and Steven Goodwin, 1998, *Looking at microbes: A microbiology laboratory for students.* Boston: Jones and Bartlett Publishers.

Patouillard, N., and A. Gaillard, 1888, Champignons du Vénézuéla et principalment de la région du Haut Orénoque récoltés en 1887 par M. A. Gaillard. *Bulletin de la Societé Mycologique de France* 4:7–46.

Polz, M. F., J. A. Ott, M. Bright, and C. M. Cavanaugh, 2000, When bacteria hitch a ride. *ASM news* 66:531–538.

Rumpho, M. E., E. J. Summer, and J. R. Meinhart, 2000, Solar-powered sea slugs. Mollusc/algal chloroplast symbiosis. *Plant Physiology* 123:29–38.

Sagan, D., and E. D. Schneider, 2000, The pleasures of change in *The forces of change.* Washington, D.C.: National Geographic Society, pp. 115–126.

Sapp, J., 1994, *Evolution by association: A history of symbiosis.* New York: Oxford University Press.

———, 2003, *Genesis: The evolution of biology.* New York: Oxford University Press.

Schaechter, M., 1997, *In the company of mushrooms*, Cambridge, Mass.: Harvard University Press.

Schneider, E. D., and J. J. Kay, 1994, Life as a manifestation of the second law of thermodynamics, *Mathematical Computer Modeling*, 19, no. 6–8:25–48.

Sciencewriter's 2003. Web page www.sciencewriters.org. Green animals and other videos of symbiosis listed on the web site.

Shapiro, James A., 1998, Bacteria as multicellular organisms. *Scientific American* 256:82–89.

Shostak, S., and V. Kolluri, 1995, Symbiotic origins of cnidarian cnidocysts. *Symbiosis* 19:1–19.

Smil, Vaclav, 2002, *The Earth's Biosphere*. Cambridge: MIT Press.

Smith, D. C., and A. E. Douglas, 1987, *The biology of symbiosis*. London: E. Arnold.

Sonea, S., and L. Mathieu, 2000, *Prokaryotology*, Montreal: Les Presses de l'Université de Montreal.

Todd, Neil B., 1970, Karyotypic fissioning and canid phylogeny. *Journal of Theoretical Biology* 26:445–480.

———, 2000, Mammalian evolution: Karyotypic Fission Theory. In L. Margulis, C. Matthews, and A. Haselton, eds., *Environmental Evolution*. second edition. Cambridge, Mass., and London: MIT Press.

———, 2001, Kinetochore reproduction underlies karyotypic fission theory: Possible legacy of symbiogenesis in mammalian chromosome evolution. *Symbiosis* 29:319–327.

Turner, J. Scott, 2000, *The extended organism: The physiology of animal-built structures*. Cambridge, Mass.: Harvard University Press.

Vernadsky, V. I., 1998, *The biosphere* (in English). Translation of Vernadsky, V. I., 1926, *The biosphere* (in Russian). New York: Copernicus Springer-Verlag.

Wallin, I. E., 1927, *Symbionticism and the origin of species*. Baltimore: Williams and Wilkins.

White, M. J. D., 1973, *Animal cytology and evolution*. Cambridge, U.K.: Cambridge University Press.

Wiener, Jonathan, 1999, *The beak of the finch*. Cambridge, Mass.: Harvard University Press.

Williamson, D. I., 1992, *Larvae and evolution: toward a new zoology*. New York: Chapman and Hall.

———2001, Larval transfer and the origin of larvae. *Zoological Journal of the Linnean Society* 131:111–122.

Wright, Barbara E., 2000, A biochemical mechanism for nonrandom mutations and evolution. *Journal of Bacteriology* 192:2993–3001.

ACKNOWLEDGMENTS

Dorion Sagan would first of all like to thank Jessica Whiteside for manuscript facilitation, feedback and help in research. He also wishes to express appreciation to Eric D. Schneider for tutelage on the thermodynamic issues. The work's largest debt is to the many scientists who have dedicated their lives to study of our examples. Only a few, for example, Rosati's *Euplotidium*, Fortey's olenids, Gupta's chimeras, Searcy's *Thermoplasma*, Bergman's *Gunnera*, Marcotrigiano's variegated leaves, Mollenhauer's *Geosiphon*, Williamson's larvae, Jeon's amebas, McFall-Ngai and Ruby's *Euprymna* squid, Heddi's weevils, Todd and Kolnicki's "karyotypic fissioning=kinetochore-reproduction" theory are mentioned by name. So many others (such as the important evolutionary work of Werner Schwemmler on insect bacteria and of Peter Atsatt on fungal-plants symbioses) are, for lack of time and space, relegated to be parts of generalizations. Some of the work of the deceased investigators such as Miehe's *Ardisia* were brought to our attention by Herr Prof. Dr. H. Linskins, University of Nejmegen, Netherlands, to whom we are grateful.

Colleagues who have helped with the manuscript in various direct and indirect ways include: David Abram, Marcus Aebi, Jennifer

Benson, Peter G. Brown, Lois Brynes, Emily Case, Michael Chapman, Michael Dolan, Ann Ferguson, June Girard, Brianne Goodspeed, Jennifer Gottlieb, Ricardo Guerrero, Jessie Gunnard, Kenneth Hsu, Michael Keston, Alan Kuzerian, Tom Kunz, Wolfgang Krumbein, James Lovelock, Adam MacConnell, Kay Mariea, Hannah Melnitsky, Vanessa Mobley, Carlos Montufar, Marta Norman, Morriss Partee, Simon Powell, Donna Reppard, Jan Sapp, Eric D. Schneider, Judith Serrin, Dean Soulia, Crispin Tickell, Gabriel Trueba, Peter Warshall, Andrew Wier, and Sean Werle. Christie Lyons and Kathryn Delisle's knowledgeable artwork, among wonderful others, grace these pages as do the works of photographers like Reg Morrison. We are especially indebted to Judith Herrick (Typro) who typed originals and revisions with great alacrity, precision, and cheerfulness. We thank Bill Frucht for his penetrating queries and insightful comments that helped shape and edit the entire work from proposal to finished book and Ernst Mayr for his superb criticism at the twenty-fifth hour. Financial help for our own studies and writing projects came from NASA Space Sciences, the Richard Lounsbery Foundation, and the University of Massachusetts Graduate School.

The hospitality of the Collegium Helveticum (Zurich) and the Hanse Wissenschaft Kolleg (Delmenhorst, Germany) was essential to the completion of this task. So was the aid of the Humboldt Foundation to which we are exceedingly grateful.

FIGURE CREDITS

FM.1 Bacterial origin of nucleated cells, drawing by Kathryn Delisle
1.1   Invisible anastomosis in beetle phogeny, drawing by Kathryn Delisle
2.1   Tree of life, drawing by Christie Lyons.
3.1   *Ardisia* leaf margin fluting due to bacterial colonies
3.2   *Euplotidium* "body farms" its defense organs, election micrographs by Prof. R. Rosati
5.1   Virus variegation: *Abutilon pictum* (Malvaceae), photos by

Michael Marcotrigiano

7.1 Scarlet cleaning shrimp *Lysmata grubhami* in green moray eel's mouth

7.2 *Pterotermes occidentis* (similar to *Heterotermes tenuis*), drawing by Laszlo Meszoly

7.3 *Pterotermes occidentis*, gut symbionts of a dry-wood-eating termite, drawing by Christa Lyons

7.4 Archaeprotists: Archaemoebae, diplomonads, parabasalids devescovinids, calonymphids, hypermastigotes, drawing by Kathryn Delisle

7.5 *Staurojoenina*: Composite individual hypermastigote, photo by David Chase

7.6 Termite city, photo by Reg Morrison

8.1 Gaia on Lovell's Earth (NASA)

9.1 *Metacoronympha, Trichonympha*, drawings by Kathryn Delisle

9.2 *Thiodendron latens* in culture at low-oxygen concentration, photo by G. A. Dubina

9.3 A karyomastigont (left) compared with an akaryomastigont (right), idealized, drawings by Kathryn Delisle

9.4 *Synderella tabogae* (many unattached nuclei and even more akaryomastigonts), photo by Michael Dolan

10.1 Williamson's "transferred larvae," drawings by Kathryn Delisle based on information from D. I. Williamson

10.2 *Hermissenda nudibranch*, a nematocyst thief, photo by Alan Kuzerian

10.3 *Euprymna scoleps*, the tiny luminous squid, photo by Margaret McFall-Ngu

10.4 Richard Fortey's trilobites: *Olenus* in their "nobody-at-home" beds, Ordovician, photo by Richard Forney

11.1 *Gunnera manicata* and its symbiotic cyanobacteriam, top photo by Louise Mead, lower right photo by Rita Kolchinski Berson

12.1 Karyotypic fission (kinetochore-centromere reproduction) theory, drawing by Kathryn Delisle, based on information from Robin Kolnicki

All illustrations not mentioned were by the authors.

INDEX